4th International Symposium on Mechanics, Structures and Materials Science (MSMS 2023)

AA004515

MATEC Web of Conferences Volume 380 (2023)

Tianjin, China
30 - 31 March 2023

Editors:

Jun Xiao
Xinwang Wu

ISBN: 978-1-7138-7341-9

Printed from e-media with permission by:

Curran Associates, Inc.
57 Morehouse Lane
Red Hook, NY 12571

Some format issues inherent in the e-media version may also appear in this print version.

For additional information, please contact EDP Sciences – Web of Conferences at the address below.

EDP Sciences – Web of Conferences
17, Avenue du Hoggar
Parc d'Activité de Courtabœuf
BP 112
F-91944 Les Ulis Cedex A
France

Phone: +33 (0) 1 69 18 75 75
Fax: +33 (0) 1 69 28 84 91

contact-edps@webofconferences.org

Additional copies of this publication are available from:

Curran Associates, Inc.
57 Morehouse Lane
Red Hook, NY 12571 USA
Phone: 845-758-0400
Fax: 845-758-2633
Email: curran@proceedings.com
Web: www.proceedings.com

TABLE OF CONTENTS

Preface: 4th International Symposium on Mechanics, Structures and Materials Science (MSMS 2023)

MSMS 2023 includes original and peer-reviewed research papers from 4th International Symposium on Mechanics, Structures and Materials Science (MSMS 2023), which held on March 30-31, 2023 in Tianjin City, China.

The mechanics, structures and materials science are experiencing considerable growth a few years ago, which involve the development of new technology requiring multidisciplinary simulation. The primary objective of this conference is to exchange knowledge among researchers, practicing engineers, technologists about the state of art in mechanics, structures, materials science and their applications. MSMS stimulates this growth by emphasizing fundamental advances that are relevant in dealing with problems of all length scales.

4th MSMS received 71 manuscripts. And 30 submissions had been accepted by our reviewers. By submitting a paper to 4th MSMS, the authors agreed to the review process and understood that papers undergo a peer-review process. Manuscripts were reviewed by appropriately qualified experts in the field selected by the Conference Committee, who gave detailed comments and-if the submission was accepted the authors would submit a revised version that took into account this feedback. All papers were reviewed using a double-blind review process: authors declared their names and affiliations in the manuscript for the reviewers to see, but reviewers did not know each other's identities, nor did the authors receive information about who had reviewed their manuscript. The Committees of 4th MSMS invested great efforts in reviewing the papers submitted to the conference and organizing the sessions to enable the participants to gain maximum benefit.

With our warmest regards,

Jun Xiao, Xinwang Wu

Conference Organizing Committee
Tianjin, China

Statement of Peer review

In submitting conference proceedings to *Web of Conferences*, the editors of the proceedings certify to the Publisher that

1. They adhere to its **Policy on Publishing Integrity** in order to safeguard good scientific practice in publishing.
2. All articles have been subjected to peer review administered by the proceedings editors.
3. Reviews have been conducted by expert referees, who have been requested to provide unbiased and constructive comments aimed, whenever possible, at improving the work.
4. Proceedings editors have taken all reasonable steps to ensure the quality of the materials they publish and their decision to accept or reject a paper for publication has been based only on the merits of the work and the relevance to the journal.

Title, date and place of the conference

4th International Symposium on Mechanics, Structures and Materials Science (MSMS 2023)
March 30-31, 2023
Tianjin, China

Proceedings editor(s):

Jun Xiao
Xinwang Wu

Date and editor's signature April 20, 2023

Research on Tangential Velocity Solution and Accuracy Analysis of Radar Target

Qi Yin, Zepeng Sun*, Yangmeng Li, Yujie Hou

Weapon Test and Training Center of Huayin, Huayin, China

Abstract. In this paper, a calculation method for the tangential velocity solution of radar target based on linear vector method is proposed. It makes full use of the advantage of high precision radial velocity measurement of radar, and realizes the accurate computation of tangential velocity of radar target. Based on the actual measurement capability of continuous wave (CW) radar, the accuracy of the tangential velocity is analyzed. The feasibility of calculation method and accuracy analysis method is verified by simulation experiment and measured data of shooting range.

Keywords: radar measurement, radial velocity, tangential velocity

1. Introduction

Radar equipment has always been playing an important role in the weapon test evaluation of range. Radar can realize real-time observation of important information such as velocity, range, azimuth and pitch angle of the target by transmitting and receiving electromagnetic waves [1]. Among many test information, radar target velocity measurement is realized based on Doppler effect [2], which indicates that the condition for the target velocity to be observed is the relative radial movement between the target and the radar, that is, the velocity measured by the radar represents the change speed of the target is moving away from or near the radar. In other words, the velocity is only a velocity component of the tangential velocity (absolute velocity) of the target in the radial direction of the radar. However, in the range test evaluation, people pay close attention to the change of tangential velocity of the target. Therefore, how to put forward an accurate calculation method of tangential velocity of target on the basis of the original observation information of radar has become a technical problem that needs to be solved urgently. In this context, a method for solving tangential velocity of radar target based on linear vector is proposed in this paper. This method makes full use of the advantage of high precision radial velocity measurement of radar, and realizes the accurate computation of tangential velocity of radar target. Simulation experiment and measured data show that the solving method is feasible.

2. Calculation method of tangential velocity

To set the projectile trajectory is shown in figure 1, $\Gamma(t)$ represents the target trajectory, the coordinate system direction is taken as the range coordinate direction, the coordinate origin O is taken as the radar coordinates (x_0, y_0, z_0), v_τ represents the ballistic tangential velocity of the target, vrrepresents the radial velocity of the target, the tangential angle ϕ represents the angle between the ballistic tangential velocity and the radial velocity of the target.

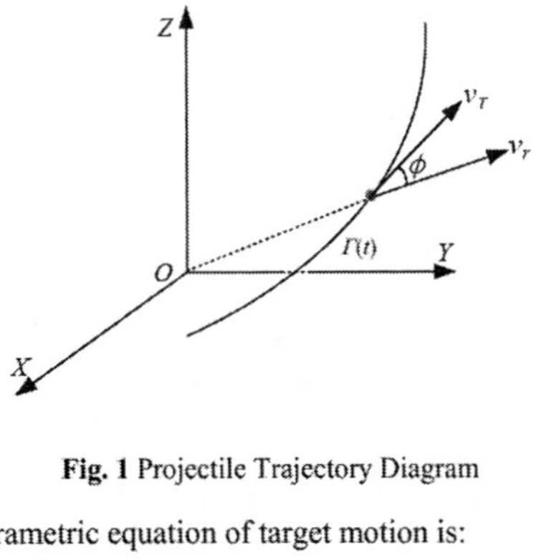

Fig. 1 Projectile Trajectory Diagram

The parametric equation of target motion is:

$$\begin{cases} x = x(t) = f_1(R,A,E) = R\cos E\cos A \\ y = y(t) = f_2(R,A,E) = R\sin E \\ z = z(t) = f_3(R,A,E) = R\cos E\sin A \end{cases} \quad (1)$$

* Corresponding author: sun_ze_peng@sina.com

As in (1), (x, y, z) represents the coordinates of the target in the range coordinate system, (f_1, f_2, f_3) represents the coordinate transformation function, and (R, A, E) represents the target range, azimuth and pitch angle measured by radar respectively, which are all functions of time.

As shown in Fig.1, the direction of radial velocity of target at a certain moment is:$(x-x_0,y-y_0,z-z_0)$, and the direction of tangential velocity is:(dx,dy,dz). According to the inner product formula of linear vectors, the tangential angle ϕ is calculated as:

$$\cos\phi=\frac{\sum_i(\psi_i-\psi_{i0})d\psi_i}{\sqrt{\sum_i(\psi_i-\psi_{i0})^2}\cdot\sqrt{\sum_i(d\psi_i)^2}} \tag{2}$$

As in (2), $\{\psi_i\}$,i=1,2,3 denotes $\{x,y,z\}$. The same as below. It should be noted that radar measurement is carried out in terms of the measurement period, which cannot be infinitely shortened due to the influence of system hardware performance and other parameters. In other words, radar output data is discrete data. Therefore, (2) needs to be discretized as follows:

$$\cos\phi=\frac{\sum_i(\psi_i-\psi_{i0})\Delta\psi_i}{\sqrt{\sum_i(\psi_i-\psi_{i0})^2}\sqrt{\sum_i(\Delta\psi_i)^2}} \tag{3}$$

As in (3),$\Delta\psi_i=\psi_i(t+\Delta t)-\psi_i(t)$

After the tangential angle between the tangential velocity and radial velocity of the target trajectory at a certain moment is solved, the tangential velocity of the target trajectory at that time can be solved by (4):

$$v_\tau=\frac{v_r}{\cos\phi} \tag{4}$$

As in (4), v_r represents the radial velocity of the target. It is the direct measurement of radar.

3. Accuracy analysis

In order to verify the feasibility of the tangential velocity calculation method proposed in this paper and research on the measurement accuracy of the method, the analysis is made as follows. First, we can take the differential from (4):

$$d_{v_\tau}=\frac{\cos\phi\cdot d_{v_r}-v_r\cdot d\cos\phi}{(\cos\phi)^2} \tag{5}$$

Eq.(5) is the tangential velocity error formula. In Formula (5), dv_τrepresents the measurement error of tangential velocity, dv_r represents the measurement error of radial velocity (depending on the actual radar capability), and d$\cos\phi$ represents the measurement error of tangential angle. In the following, the measurement error of tangential angle is analyzed, and the variable substitution is as follows:

$$\alpha=\sum_i(\psi_i-\psi_{i0})\Delta\psi_i$$

$$\beta=\sqrt{\sum_i(\psi_i-\psi_{i0})^2}$$

$$\gamma=\sqrt{\sum_i(\Delta\psi_i)^2}$$

Then, we can take the differential from (3):

$$d\cos\phi=\frac{d\alpha\cdot(\beta\cdot\gamma)-\alpha\cdot(\beta d\gamma-\gamma d\beta)}{(\beta\cdot\gamma)^2} \tag{6}$$

According to (6), the measurement error of tangential angle is related to the coordinate change error $(d\Delta\psi_i)$ and the coordinate error$(d\psi_i)$. Then, the analysis is made as follows: In a measurement period of radar, the target trajectory$\Gamma(t)$ is continuous function of time and has continuous partial derivative, according to Lagrange's mean value theorem [3]:

$$\Delta\psi_i=\sum_i\frac{\partial f_i}{\partial\xi_i}(R,A,E)\Delta\xi_i$$

In the last equation, $\{\xi_i\}$, i=1,2,3 denotes $\{R, A, E\}$.

Take the differential from the last equation:

$$d\Delta\psi_i=\sum_j\sum_i\left(\frac{\partial^2 f_i}{\partial\xi_j\partial\xi_i}(R,A,E)d\xi_i\right)\Delta\xi_j + \frac{\partial f_j}{\partial\xi_j}(R,A,E)\cdot d\Delta\xi_j \tag{7}$$

As in (7),$\{\xi_i\}$,$\{\xi_j\}$,i,j=1,2,3 denotes $\{R,A,E\}$, $\{d\xi_i\}$ denotes $\{dR,dA,dE\}$, which are the measurement error of range, azimuth and pitch angle of radar, depending on the actual capability of radar. $\{d\Delta\xi_j\}$ denotes $\{d\Delta R,d\Delta A,d\Delta E\}$, which are the measurement errors of range, azimuth and pitch angle variation of radar. These measurement errors are high-order small quantities of range error and angle measurement error of radar. The measurement errors of radar range variation and angle variation are analyzed as follows:

According to Lagrange's mean value theorem, the variation of radar range measurement can be expressed as:

$$\begin{aligned}\Delta R&=R(t+\Delta t)-R(t)\\&=R'(t+\theta\Delta t)\cdot\Delta t=\tilde{\nu}_r(t)\cdot\Delta t\end{aligned}$$

In the last equation, $\tilde{\nu}_r$ represents the radial velocity of the target radial distance differentiated with respect to time. Then, we can take the differential from above equation:

$$d\Delta R=d\tilde{\nu}_r\cdot\Delta t$$

As in aboveequation, $d\tilde{\nu}_r$ represents the measurement error of radial velocity of differential method. According to error theory [4]:

$$d\tilde{\nu}_r=\sqrt{\left(\frac{\Delta R}{\Delta t}-v_r\right)^2+(dv_r)^2}$$

In the last equation, v_r represents the radial velocity measured by radar, which is obtained through frequency measurement. $v_r$has high measurement precision. dv_rrepresents the measurement error of the velocity.

Similarly, the measurement error of radar angle variation can be expressed as:

$$d\Delta A = d\tilde{\omega}_A \Delta t$$
$$d\Delta E = d\tilde{\omega}_E \Delta t$$

As in aboveequation, $\tilde{\omega}_A$ and $\tilde{\omega}_E$ respectively represent the target azimuth and pitch angle velocity obtained by differentiating time. It should be noted that target azimuth and pitch angle velocity are not directly measured by radar, and the measurement errors of azimuth and pitch angle velocity are not given in radar technical index. Therefore, the measurement errors of $\tilde{\omega}_A$ and $\tilde{\omega}_E$ cannot be directly calculated by error theory. In order to solve this problem, the minimum mean square error optimization algorithm is adopted to estimate the measurement errors of $\tilde{\omega}_A$ and $\tilde{\omega}_E$. (details below).

On the other hand, radar coordinate measurement errors (dx,dy,dz) can be expressed as follows according to the coordinate transformation formula:

$$d\psi_i = \sum_i \frac{\partial f_i}{\partial \xi_i}(R,A,E)d\xi_i \quad (8)$$

Therefore, after the coordinate variationmeasurement error $d\Delta\psi_i$ and coordinate measurement error $d\psi_i$ are defined, the measurement error of tangential angle can be solved by (6). And then the measurement error of tangential velocity of radar target trajectory can be solved by (5).

4. Test Verification

In order to verify the feasibility of the tangential velocity calculation method and precision analysis presented in this paper; Firstly, the ballistic simulation software of the range is used to simulate the trajectory of the projectile. Based on the projectile firing parameters and artillery firing data in the performance evaluation test of a projectile, the simulation test is carried out. The radial and tangential velocity of the projectile obtained are shown in Fig. 2. The tangential velocity (solid red line in Fig. 2) obtained from the simulation test is taken as the truth-value to verify the feasibility of the tangential velocity accuracy analysis method proposed in this paper. The measurement errors of $\tilde{\omega}_A$ and $\tilde{\omega}_E$ are ignored when verifying the accuracy analysis method of the tangential velocity. The test results are shown in Fig 3.

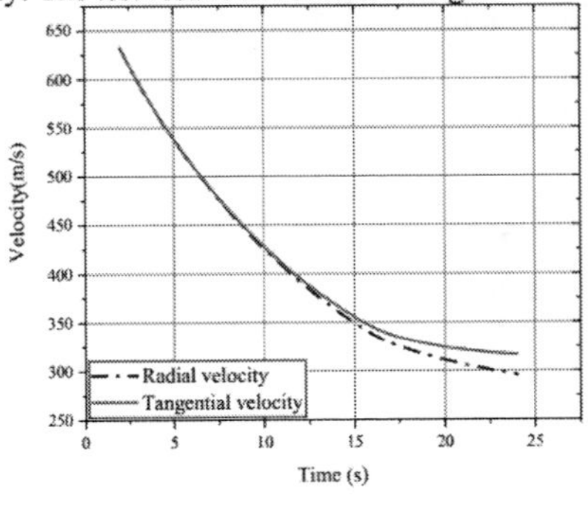

Fig. 2. Velocity curve of a projectile

As shown in Fig.3, the solid red line in the figure represents the truth-value of tangential velocity measurement error, which is the result of the difference between the tangential velocity calculation value obtained by linear vector method and the truth-value of tangential velocity obtained by simulation test (solid red line in Fig.2).

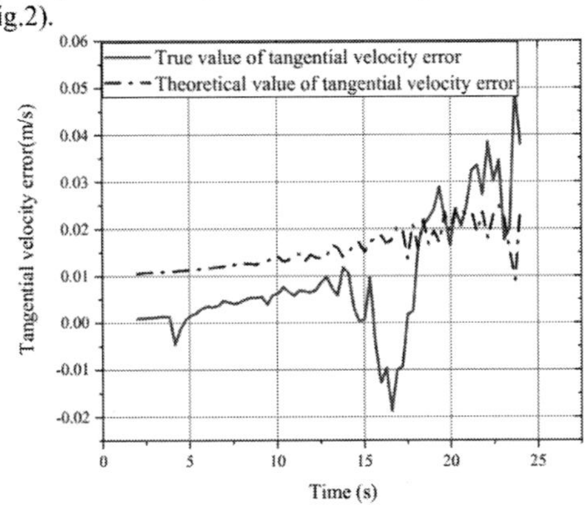

Fig. 3. Tangential velocity error curve of a projectile

The blue dashed line in the figure represents the theoretical value of tangential velocity error, which is the calculation result of the accuracy analysis method proposed in this paper. The root-mean-square value of tangential velocity error was statistically analyzed under the two conditions. The root-mean-square value of tangential velocity error represented by the solid red line is 0.0185. The blue dashed line shows that the root-mean-square value of tangential velocity error is 0.0160. Under these two conditions, the trends of red and blue curves are quite different. This is because of the measurement errors of $\tilde{\omega}_A$ and $\tilde{\omega}_E$ are ignored. To solve this problem, the minimum mean square error optimization algorithm is adopted to estimate the measurement errors of $\tilde{\omega}_A$ and $\tilde{\omega}_E$.

In Fig.3, the red curve represents the truth-value of tangential velocity measurement error. The objective function can be expressed as follows:

$$Q = \sum_{i=1}^{n}(Z_i - X_i d\tilde{\omega}_A - Y_i d\tilde{\omega}_E)^2 \quad (9)$$

As in (9),X_i,Y_i,Z_i represent measurement matrices, which can be derived from (6), (7) and(8). To minimize the value of the objective function:

$$d\tilde{\omega}_A = \frac{\begin{vmatrix} \sum_i X_i Z_i & \sum_i X_i Y_i \\ \sum_i Y_i Z_i & \sum_i Y_i^2 \end{vmatrix}}{\begin{vmatrix} \sum_i X_i^2 & \sum_i X_i Y_i \\ \sum_i X_i Y_i & \sum_i Y_i^2 \end{vmatrix}}$$

$$d\tilde{\omega}_E = \frac{\begin{vmatrix} \sum_i X_i^2 & \sum_i X_i Z_i \\ \sum_i X_i Y_i & \sum_i Y_i Z_i \end{vmatrix}}{\begin{vmatrix} \sum_i X_i^2 & \sum_i X_i Y_i \\ \sum_i X_i Y_i & \sum_i Y_i^2 \end{vmatrix}}$$

It can be obtained by substituting the numerical value: $d\tilde{\omega}_A$=0.0072324mrad/s, $d\tilde{\omega}_E$=-0.000051018mrad/s. The theoretical value of tangential velocity error is recalculated, and the calculation result is plotted in Fig.4.

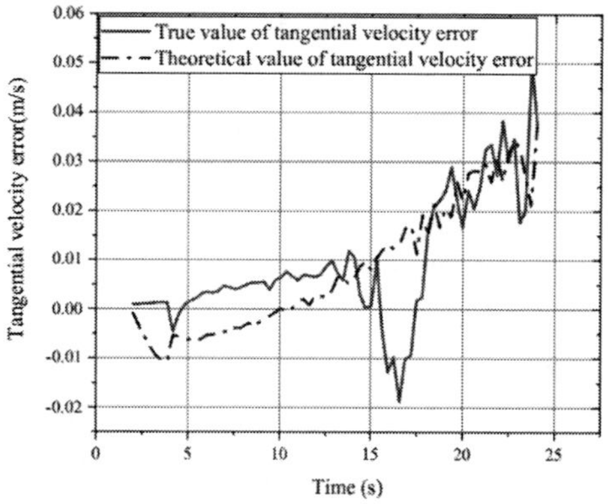

Fig. 4. Tangential velocity error curve of a projectile

By comparing Fig.3 and Fig4, it can be concluded that the tangential velocity error has better trend stability between the theoretical value (blue dashed line in the figure) and the truth-value (solid red line in the figure) under the condition of considering the measurement errors of $\tilde{\omega}_A$ and $\tilde{\omega}_E$. The root-mean-square value of the tangential velocity theoretical error under this condition is 0.0165, which is closer to the root-mean-square value (0.0185) of the tangential velocity measurement error under the actual condition. Therefore, it is feasible to adopt the minimum mean square error optimization algorithm to estimate the measurement errors of $\tilde{\omega}_A$ and $\tilde{\omega}_E$.

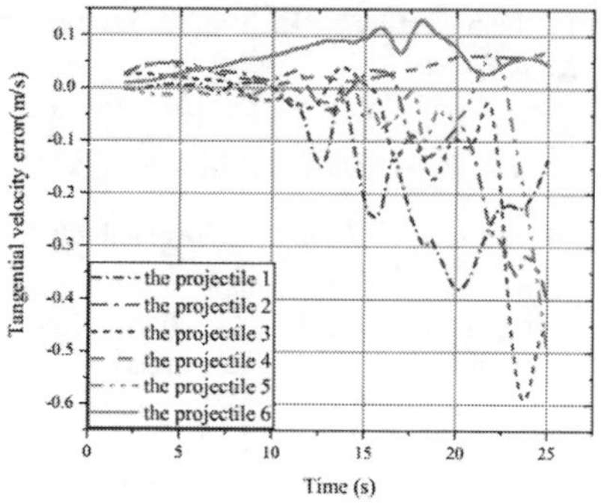

Fig. 5. Tangential velocity error curves of various projectile types

Furthermore, to verify the generality of the conclusion, it is necessary to verify that the precision obtained by the analysis method proposed in this paper does not diverge with the change of projectile type. Therefore, the tangential velocity errors of several different types of projectiles are analyzed in the same observation time. The error curves are plotted in Fig.5, and the analysis results are listed in Table I. It can be concluded from Table I that during the observation time, for different types of projectiles, the root-mean-square value of tangential velocity measurement error based on the linear vector method does not exceed 0.2m/s, and does not diverge because of different types of projectiles. In addition, it should be noted that the fluctuations of measured data are relatively large compared with the simulation data, so the error analysis results based on the measured data are relatively large compared with the simulation results. On the other hand, due to the different shooting conditions of the projectiles, there is a deviation between the velocity errors of different projectiles, which is in line with the general rule.

Table 1. Comparison of precision analysis

Test error / Type of projectile	Tangential velocity measurement error (m/s)
The projectile 1	0.179
The projectile 2	0.136
The projectile 3	0.175
The projectile 4	0.041
The projectile 5	0.093
The projectile 6	0.069

Acknowledgments

This work was financially supported by national defense research fund.

References

1. G. Eason, B. Noble, and I. N. Sneddon, "On certain integrals of Lipschitz-Hankel type involving products of Bessel functions," Phil. Trans. Roy. Soc. London, vol. A247, pp. 529–551, April 1955.
2. J. Clerk Maxwell, A Treatise on Electricity and Magnetism, 3rd ed., vol. 2. Oxford: Clarendon, 1892, pp.68–73.

3. I. S. Jacobs and C. P. Bean, "Fine particles, thin films and exchange anisotropy," in Magnetism, vol. III, G. T. Rado and H. Suhl, Eds. New York: Academic, 1963, pp. 271–350.
4. K. Elissa, "Title of paper if known," unpublished.
5. R. Nicole, "Title of paper with only first word capitalized," J. Name Stand. Abbrev., in press.
6. Y. Yorozu, M. Hirano, K. Oka, and Y. Tagawa, "Electron spectroscopy studies on magneto-optical media and plastic substrate interface," IEEE Transl. J. Magn. Japan, vol. 2, pp. 740–741, August 1987 [Digests 9th Annual Conf. Magnetics Japan, p. 301, 1982].
7. M. Young, The Technical Writer's Handbook. Mill Valley, CA: University Science, 1989.

Design of Parts Flexible Palletizing System

Yuan Yang[1], Ligang Qiang[2], Yong Zhang[2], Xudong Yang[1]

[1]School of mechanical engineering, Guizhou University, Guiyang 550000, China
[2]Aerospace Jiangnan Group Co., Ltd, Guiyang 550000, China

Abstract. Part palletizing packaging is an important part of the logistics industry. Traditional manual palletizing can no longer meet today 's order volume. In order to improve the production efficiency of the palletizing process, a part flexible palletizing system was designed. The S7-1200 PLC was used as the controller, and the high-precision altimeter and length measurement photoelectric sensors were used to check the parts in real time, and the manipulator and the end actuator were used for grasping and stacking. The end actuator with multi-axis synchronous control technology is adopted to realize simultaneous grasping of multiple boxes of different specifications, which greatly improves the production efficiency and automation level of part palletizing. This flexible palletizing system and robotic gripping technology greatly improves the efficiency of gripping small objects and also allows for extended applications in the field of FAST disassembly and handling robots.

Key Words: flexible palletizing; PLC; manipulator end actuator; manipulator.

1. Introduction

Material sorting and palletizing is a common process in manufacturing, logistics and transportation industries. In the past, the sorting and palletizing of materials were completed manually, which not only had high labor intensity, but also had a high error rate of sorting and palletizing, which made the process inefficient12. In order to improve efficiency, the targeted technical transformation was needed to conform to the development trend of automation. The different specifications of the same material also increase, which brings great challenges to the palletizing of small materials34. Based on this background, the part palletizing system is designed. PLC is used as the controller, combined with the length and height sensor to realize the detection and palletizing of parts, and the multi-axis palletizing mechanism is controlled to realize the simultaneous grabbing of multiple specifications, which accelerates the palletizing efficiency. PLC through Ethernet connection touch screen, PC and manipulator controller, to achieve human-computer interaction, equipment monitoring and manipulator movement, thus improving the automation level of palletizing process.

2. Part Flexible Palletizing System

2.1 Structure of Part Flexible Palletizing System

The flexible palletizing system mainly consists of grab belt, manipulator, manipulator end actuator, pallet and control cabinet, as shown in figure 1.The grabbing belt is mainly powered by a servo motor, and the belt is equipped with a corresponding in-place photoelectric sensor and a length-measuring photoelectric sensor to realize the positioning and review of the part 5. The system uses an EPSON LS20-B804S robot with the advantages of high speed and high performance ratio. The end actuator is the key to realizing the flexible palletizing system, using 9 stepper motors to control 9 shafts, to achieve the suction cup in the X and Z direction of movement. Pallets are the necessary structure from palletizing to packaging. The control cabinet is the control center of the whole system, which contains PLC, touch screen and other electrical components, completing the system control, communication and human-computer interaction.

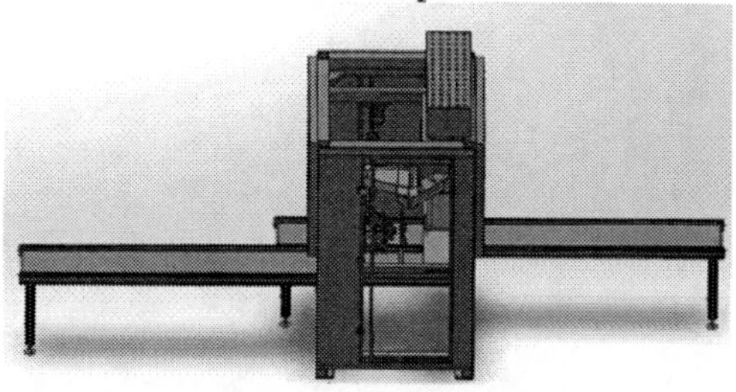

Fig. 1 Part Flexible Palletizing System

2.2 Part Flexible Palletizing System Process

The sorted parts are transported to the top of the grasping belt through the cache belt, and the parts are reviewed by the photoelectric sensor and the length measuring

 https://doi.org/10.1051/matecconf/202338001002

photoelectric sensor. After the review is completed, the grasping signal is sent. After the manipulator waiting above the grabbing point receives the grabbing signal, the

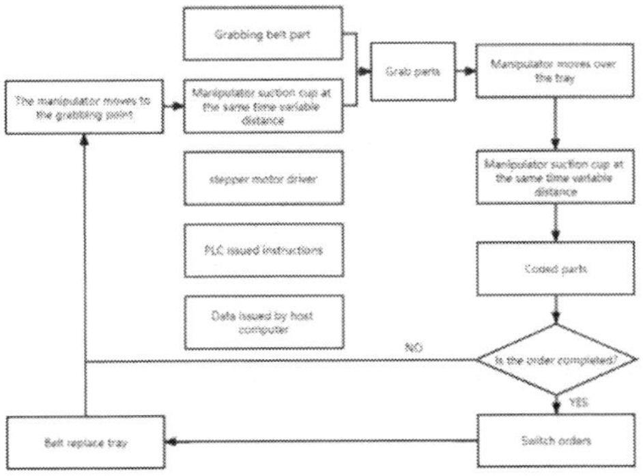

Fig. 2 Process Flow

suction cup grabs the part with variable distance. When the tray reaches below the code point, it sends an allowable dispensing signal to the manipulator. After receiving the part release signal, the manipulator moves to the parts release point for parts release. After the part is released, the manipulator returns to the above of the grasping point and waits for the instruction. This process is repeated in the order after the completion of the part palletizing signal, the tray is delivered to the packaging station by belt, as shown in Figure 2.

3. Control Hardware Design.

The hardware of the part flexible palletizing system was designed according to the process flow. The system mainly includes palletizing mechanism and electric control cabinet. Palletizing mechanism mainly uses the EPSON LS20-B804S type manipulator, its structure also includes the manipulator end actuator, tray, grab belt and tray belt. Among them, 7 groups of position photoelectric sensors and a group of length measuring photoelectric sensors are used. One group of position photoelectric sensors is used to judge the position of the tray, the other 5 groups are used to judge the position of the part, and the last group is the limit position of the belt. The end actuator of the manipulator is the key technology of this paper. It is equipped with 9 stepping motors, which provides X and Z axis power for the sucker. Because the end actuator is based on the third set of suckers, four origin slot photoelectric sensors are also set up to determine the X-axis origin of the 1,2,4 and 5 sets of suckers. At the same time, five altimeter photoelectric sensors were set up to review the part specifications. The end actuator of the manipulator is shown in Figure 3. The vacuum pump, solenoid valve, three-color signal lamp and buzzer are also set outside. The cooperation of the vacuum pump and the solenoid valve realizes the suction and blowing of the suction cup, and ensures the grasping and discharge of the part.

Fig. 3 Manipulator End Actuator

The electric control cabinet is the hardware core of the part of system, which can be divided into the terminal plug, insurance switch, fuse, transformer, servo motor controller, stepper motor driver, relay, power supply, control terminal, switch, touch screen and manipulator controller, as shown below. The terminal socket is used to transfer the power line and signal line inside and outside the cabinet, and the insurance switch controls the on-off power of the electrical components outside the cabinet. The fuse is the insurance part of the control cabinet to ensure that the circuit can be protected in the case of short circuit. The transformer converts 380V voltage to 24V. Servo motor controller is to control the grab belt and tray belt start and stop and anyway, stepper motor drive by changing the pulse and direction control suction movement. The control end includes three SIMATIC S7-1200 PLCs, one of which is the main control PLC, which is responsible for the control and calculation of the whole system. The other two are XPLC and ZPLC, which control the X-axis motion and Z-axis motion of the sucker in the end actuator of the manipulator respectively. The purpose is to share the calculation amount of the main control PLC and improve the operation speed. The touch screen is the main interface of human-computer interaction. It can not only observe the control information, but also control the operation of the end actuator of the manipulator. Manipulator controller is the brain of the manipulator, which controls the movement of the manipulator. The controller of EPSON RC90 is used to achieve the effect of fast communication through the connection with PLC Ethernet. The layout is shown in Figure 4.

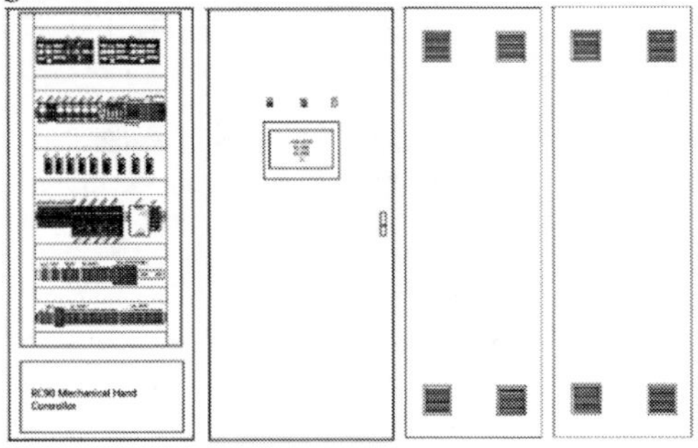

Fig. 4 Control Cabinet

The combination of the palletizing mechanism and the control cabinet realizes the whole process of the system from grasping the part to palletizing. The coordinated motion of the end actuator of the manipulator is realized by multi-axis synchronous control technology, which ensures the synchronization of the end actuator. The hardware structure of the whole system is shown in Fig.5.

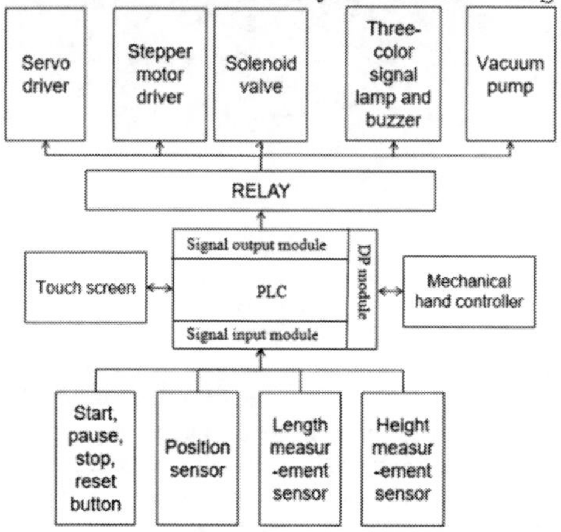

Fig. 5 Hardware Structure Chart

4. Control Software Design

The hardware system is the carrier of the control system, and the software system is the key to measure the stability and reliability of the whole system. The part flexible palletizing system software is not only required to realize the control of the movement of each mechanism, but also to monitor the operation of the equipment in real time. It has the functions of alarming and protecting the safety of workers, and can provide a convenient human-computer interaction system.

4.1 PLC Program

The main control PLC receives the palletizing data sent by the host computer, and then after the system is started, the three PLCs enable the manipulator to reset the X and Z axes, and detect the zero return of each axis through the slot photoelectric sensor; after the reset is completed, wait for the write of each axis moving data; the manipulator receives the standby point signal and moves to the standby point while the flexible sucker changes the distance. The height and length of the part are then checked by the height sensor and the length sensor; after the review is correct, the part signal is issued to the manipulator to grab the part; at the same time, the PLC performs the drop detection. If it is correct, the signal above the dispensing point is sent to the manipulator and moved to the top of the dispensing point. At the same time, the PLC allows the palletizing signal to be returned. When the manipulator receives the allowable palletizing signal, the solenoid valve is started, and the flexible sucker is changed to break the vacuum dispensing. After the part is released, the manipulator again receives the standby point signal and returns to the standby point to cycle this process. Fig.6 is the control flow chart.

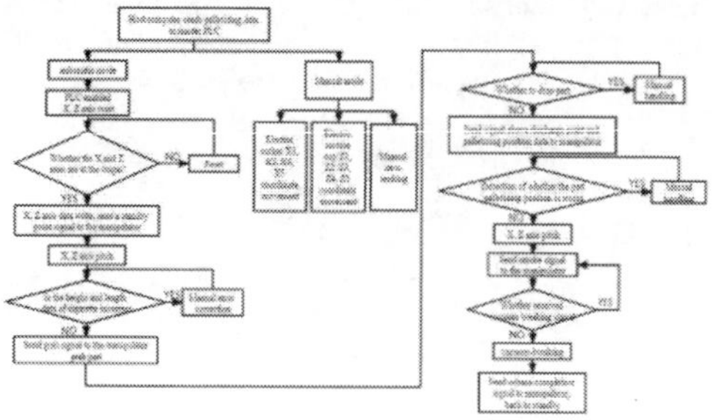

Fig. 6 The Control Flow Chart

4.2 Touch Screen Design

Touch screen mainly realizes the equipment running state monitoring, palletizing data manual application, manual click device, fault display, clear data, start the suspension system and other functions. The touch screen has six interfaces, namely Palletizing Main Interface, Sensor Parameters, Data Interaction, Palletizing Data, Sucker Motor and Manipulator. The bit state switching switch in the interface is directly connected to the relevant contacts in the electrical control program through the connection port Net, select the corresponding address type and then enter the corresponding contact address for communication. Achieve the control of each button on the system. When the flexible sucker reaches the zero point, the electricity will automatically stop. This method is to obtain the signal of the pause shaft by connecting a zero-point switch in series with a positive and negative normally open contact.

The main interface of palletizing mainly includes five parts, Virtual Palletizing, Alarm Message, Normal Operation, Palletizing Information and interface switching button. The virtual palletizing simulates the palletizing process, and can observe the signals of 6 sets of in-place photoelectric sensors and 1 set of length measuring photoelectric sensors. It also simulates the position of the end actuator and the order number of the tray. This part can also perform the belt 's point movement, buzzer and safety door status. Alarm Message is used to display the alarm information of the system, so that the operator can more accurately find the location of the failure, in order to solve faster. The Normal Operation section is set up START, STEP, RESET, Vacuum-Breaking, Manual Grab, Manual Palletizing, Manual Vacuum Pump and Total Clear Data. Each button controls the corresponding functions, among which Manual Grab and Manual Palletizing control one layer and one single manual palletizing respectively. Palletizing Information displays Odd Number, Coded, Overall Amount and Height. The interface switching button controls the switching of various interfaces, including Sensor Parameters, Data Interaction, Palletizing Data, Sucker Motor and Manipulator.

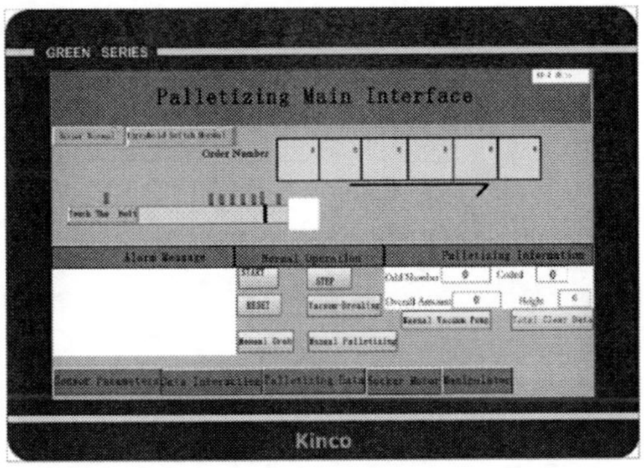

Fig. 7 Palletizing Main Interface

5. Conclusion

In order to improve the palletizing efficiency of the part logistics center, this paper designs a set of flexible part palletizing system, which breaks the limitation that the manipulator can only grab a single material, and realizes the simultaneous grabbing of multiple parts. The system takes SIEMENS S7-1200 PLC as the control core, and reduces the amount of computation of a single PLC by using a master-slave PLC. The multi-axis synchronous control technology is used to control the motor motion of the end actuator, and the synchronization of each sucker is realized. The touch screen is designed again, which can display the motion of the sucker, alarm information and palletizing data, and realize the human-computer interaction function. This system improves the automation level of the part palletizing system and improves the efficiency of palletizing. The flexible palletizing system and robotic gripping technology significantly improves the efficiency of small object gripping, and the technology provides a good theoretical and applied research basis for the group's next research on the Five-hundred-meter Aperture Spherical radio Telescope (FAST) feeder receiver disassembly and handling robot system, which further enables maintenance operation robots to efficiently handle the parts of the FAST feeder receiver that require maintenance.

Acknowledgements

This work was financially supported by Guizhou Tobacco Corporation Science and Technology Project (Research on Optimization of Sorting Scheduling and Delivery Service Based on Business Flow Data Driven (No. 2022)) and National Key Research and Development Plan (Feeder Receiver Disassembly Robot and Wireless Interference Intelligent Detection System [2019YFB1312704]).

References

1. Chunsheng Hu, Hongxing Wei, Xiaopeng Yan, Guoli Li. Research and Application of Palletizing Robot [J]. Computer Engineering and Application, 2022, 58 (02): 57-77.
2. Xiangao Wang, Jianyong Guo, Peng Su, Yaohui Rong, Qianglong Li. Design of material sorting and palletizing monitoring system based on internet of things and PLC [J]. Journal of Hengshui University, 2022,24 (04): 11-16 + 28.
3. Juanlan Qu, Xiaokang Chen, Quanfang Zhang, Tianhao Zhang, Shaowei Wang. Simulation design of palletizing robot workstation based on RobotStudio [J]. Southern Agricultural Machinery, 2022,53 (20): 19-21 + 28.
4. Yangyang Tang.Design of automatic sorting and stacking system for non-standard wood boards[J]. Lo Technology,2022,41(05):125-127.
5. Mingyang Xu, Xudong Yang, Ruiqiang Pu, Yuanfu Hu. Design of Automatic Palletizing Control System for Profiled Cigarette [J]. Packaging Engineering, 2020, 41 (23): 187-194.

Cause Analysis and Process Optimization of Non uniformity of FRP Anti current Vibration Carrier Shell

Yanqiong Liu, Ji Wu, Peng Hou, Wenguan Wang, Lijie Men

Dalian Scientific Test & Control Technology Institute, Dalian116013, China

Abstract. For improving the test system, it needs to build measuring carrier and optimize shell design. To obtain valid and reliable test data, it needs the measurement uncertainty of test system with 3dB.But the test results show the measurement of results between the inside and outside the shell are quite different. This has great impact on the measurement results. This paper carried out cause analysis from five links of people, machine, material, method, environment for sound transmission performance of FRP housing. It found out the main reason of poor sound transmission of FRP housing, Single side vacuum adsorption molding method have been taken, good results and expected purpose have been achieved.

Key word: sound transmission, FRP shell, Single side vacuum adsorption molding method.

1. Introduction

In the underwater radiated noise test system, in order to reduce the interference of the ocean background to the test system, improve the measurement capability and the validity and reliability of the measurement data, It is necessary to build a measuring carrier with anti current and vibration, Optimize the design of hydrophone shell carrier, improve the anti current and vibration performance of the test system. Related research at home and abroad found that FRP materials have light weight, high strength, chemical corrosion resistance, good electrical insulation and sound transmission etc. It can be used as the shell material of anti current and vibration damping carrier. It is found that the thickness of anti current vibration carrier is uneven, to ensure the measurement uncertainty of the test system is within 3dB, Improve the Measurement Reliability of FRP Anti current Vibration Absorbing Carrier, In this paper, the influence of the thickness and non-uniformity of FRP on the sound transmission performance is studied, Optimize the structure design and process flow, improve the yield of structural carrier, and then improve the sound transmission performance of anti current vibration carrier.

2. Analysis of sound transmission of FRP shell

2.1 Anti current vibration carrier

Anti current vibration carrier shall meet the following requirements: (1) It has good resistance to current and vibration, and it can effectively reduce the influence of self noise of hydrophone.(2) It has good sound transmission performance to ensure the undistorted transmission of acoustic signals to the maximum extent;(3) It can fix the position of hydrophone; (4) It can prevent the hydrophone from being damaged by marine organisms. The schematic diagram of underwater acoustic signal transmission through the anti current damping vibration is as follows.

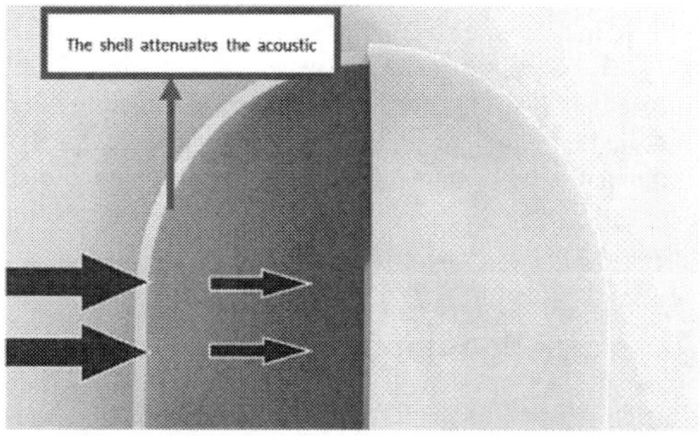

Fig. 1 Schematic Diagram of Acoustic Signal Passing through FRP Shell

2.2 Analysis of the influence of thickness on sound transmission performance of FRP

To study the influence of thickness on sound transmission performance of FRP, FRP test pieces of the same material are selected for this sample, it selects 3mm, 4mm, 5mm and 6mm thick materials as samples, tested by professional metrology institution, the comparison of sound transmission coefficients of FRP shells with different thicknesses is shown in Figure 2.

The following conclusions are drawn from the investigation of the influence of the thickness on the sound transmission performance of the FRP shell: The thickness of each part of FRP housing is uneven, thickness varies from 3mm-7mm. For broadband noise, the smaller the thickness of FRP is, the higher the sound transmission coefficient is. The sound transmission coefficient value bit of 6mm thick is 0.89, the sound transmission coefficient value bit of 3mm thick is 0.98.

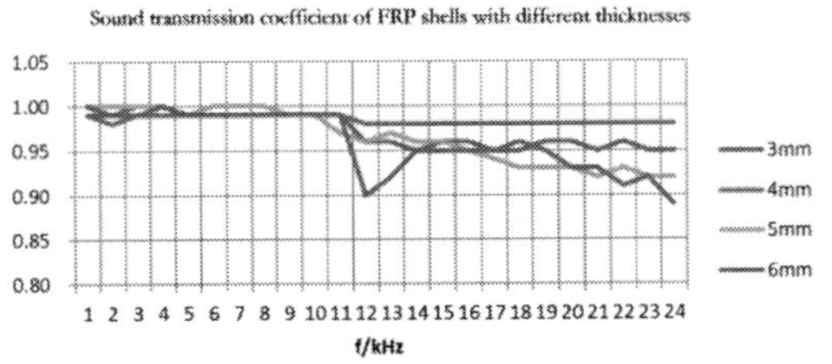

Fig. 2 Sound Transmission Coefficient of FRP Shell Materials with Different Thickness

3. Cause Analysis and Process Optimization of Uneven Shell Thickness

3.1 Analysis of the cause of non-uniformity of shell thickness

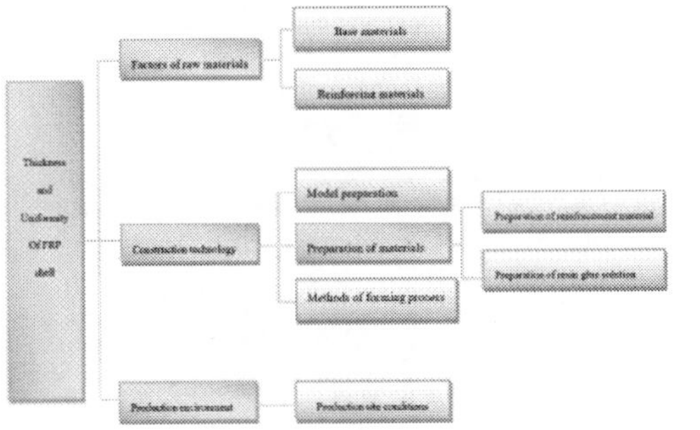

Fig. 3 the cause tree diagram of the thickness and uniformity of the FRP shell

Table 1. Analysis and Statistics of Uneven Causes

No	End cause	Analysis content	Confirm the standard/impact degree	Impact Analysis
1	Base materials	Whether the selected base material affects the shell thickness and uniformity	The matrix material is epoxy resin, which has good mechanical properties, high toughness, good corrosion resistance, strong wettability, strong adhesion with fibers, and small shrinkage.	Imperatives
2	Reinforcing materials	Whether the selected reinforcement material affects the shell thickness and uniformity	The material has a fine diameter, excellent resin permeability, and a thickness of 0.1mm-0.5mm.	Imperatives
3	Mold preparation	Check whether the mold preparation conforms to the operating procedures	The mold is complete, clean, coated with release agent, and the mold is clean without residue, which has little impact on the uniformity of the shell.	Imperatives
4	Preparation of reinforcement materials	Check whether the reinforcement material is prepared in accordance with the operating procedures	Use materials that meet specifications. The cutting size meets the requirements; Materials are identified.	Imperatives
5	Resin liquid preparation	Check whether the preparation of resin solution conforms to the operating procedures	The resin viscosity is 125 ± 10mpa. s, and each process is carried out according to the operation process, which will not affect the uniformity and thickness of the FRP shell.	Imperatives
6	Molding process	Whether the thickness and uniformity of the shell are affected by the molding process	The resin is smeared with simple and rough tools such as scraper, brush and scissors, which is easily affected by the skill level of operators and is difficult to smear the resin evenly	cause
7	Production site conditions	Standardization of production site conditions	The site environment is clean and tidy, the temperature is 25 ℃ ± 2 ℃, and the humidity is 50% - 60%, it meets the production requirements.	Imperatives

For the non-uniformity of shell thickness, the causes of poor sound transmission of FRP are analyzed from five aspects: man, machine, material, method and environment, it draws the cause tree diagram of the thickness and uniformity of the FRP shell, It analyzes the causes of 7 end causes respectively, the main reasons for uneven shell thickness are obtained, as shown in Table 1.

3.2 Causal analysis

Hand pasting method is adopted for initial production of FRP shell, the process flow is shown in Figure 4. When the glass fiber reinforced plastic shell is made by hand pasting method, it uses scraper, brush, scissors and other simple and rough tools to apply resin, it is difficult to apply the resin evenly. At the same time, the thickness and evenness of each part are also unstable, which is easily affected by the skill level of operators, so the stability of product quality is poor. Due to the complexity of the shell, there are many curved surfaces and radians, the groove corners are not fully impregnated with resin by hand pasting process, resulting in dislocation, displacement and uneven distribution of glass fibers, even resulting in uneven thickness of each part.

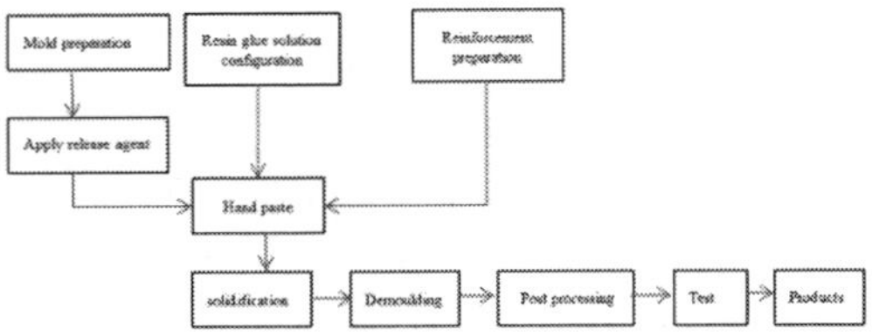

Figure 4. Hand pasting process flow chart

For the shell processed by hand pasting process, the thinnest part is 3mm, the thickest part is 7mm, and the thicker part is mostly found at the corners with radians. After many tests, the shell thickness has been improved to 5mm, but it is still difficult to reach the target thickness of 3mm.

3.3 Process optimization

The hand paste process is greatly affected by human factors, it cannot reach the target thickness. It learns from advanced process experience at home and abroad, and combines with the structural characteristics of the anti current and vibration carrier, three optimization processes are proposed, including molding scheme, molding vacuum forming negative pressure method, and single-sided vacuum adsorption molding method, the scheme is analyzed and verified to determine the optimal one.

3.3.1 Molding scheme

The molding method in the molding process is adopted .the metal molding is used to constrain the molding material between the molding surfaces of the two halves of the mold. The molding material is solidified under a certain temperature and pressure.

The molding method adopted in this scheme makes the surface roughness of the test housing reach the ideal level, However, as the filling material and staple fiber are clamped by the mold during molding, the internal situation is invisible, if the problem cannot be solved in time, the void ratio and thickness in this scheme cannot meet the test requirements, and the scheme needs to be further improved.

3.3.2 Method of forming negative pressure by mould pressing and vacuum pumping

Lay glass fiber, mold release agent, filling materials, etc. on the cured gel coat, and seal the edges around the cavity tightly. Vacuum the cavity and inject resin into the cavity. Under vacuum ,the resin flows along the resin pipeline and the fiber surface flows to impregnate the fiber bundle, so as to achieve overall pressure, and the product is solidified and shaped under room temperature or heating conditions.
Scheme features: It has improved mechanical properties, The strength, stiffness and other physical properties of the components formed by vacuum induction process can be increased by 1.5 times; It has good repeatability, relatively constant resin ratio of components, low porosity, generally ≤ 1%, which is greatly improved compared with hand pasting process; It is light in weight, with fiber content up to 75-80%, and requires no additional materials to connect the core materials; Environmentally friendly, the vacuum induction process is almost a closed mold forming process, and volatile organic compounds and toxic air pollutants are limited in the vacuum bag.

This scheme has outstanding advantages and is very suitable for the manufacturing of test shells. However, due to the lack of manual intervention, some indicators still need to be improved. This method requires upgrading of existing equipment, professional training of operators, and improvement of materials and manufacturing methods. Due to high cost of time and capital, it cannot be completed in a short period of time, so it still needs to be improved on the basis of this scheme.

3.3.3 Single side vacuum adsorption molding method

The method of forming negative pressure by mold pressing and vacuum pumping is improved, it's a liquid pressure method, the process advantages of scheme II are retained, and the production process of carbon fiber is used for reference, the outer surface of the shell is vacuumized on one side, and the inner surface uses a vacuum bag to adsorb the material onto the shell, at the same time, manual intervention shall be carried out during the curing process. Once problems are found, timely intervention shall be carried out to finally cure and shape.
Scheme features: The time cost and capital cost are greatly reduced, and the shell required for the test can be manufactured.
Through group discussion and comparison, the first method is too rough, and the sound transmission performance of the FRP shell produced by processing is not improved greatly, which still cannot meet our test requirements; The improvement effect of Scheme II is the best. However, due to the need to purchase expensive equipment, the high capital cost and time cost of investment, and the lack of human intervention in the processing of glass fiber reinforced plastic shell, it is easy to produce defective products, so it is not considered. Finally, the team members agreed to adopt Scheme III - single-sided vacuum adsorption molding method, which is easy to operate and has low investment cost. The thickness of the FRP shell can be controlled within 3mm, and the shell surface is also very uniform.

3.4 Effect verification

Single side vacuum adsorption molding method is adopted for construction, the effect of samples is checked, the measuring points at typical positions on the shell surface are selected for measurement, the thickness of each position of the improved process is shown in the figure below,
After implementing the new scheme, the gap disappears, the thickness is between 2.8mm-3mm, the deformation is small, the roughness is about 3.2 μ m, the sound transmission coefficient is 0.98-0.99, the sound transmission performance is excellent, which meets the technical requirements, and can be used as the construction process of anti current and vibration reduction carrier.

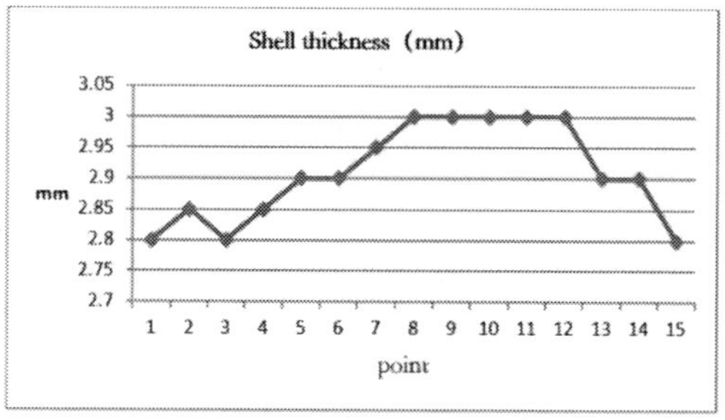

Figure 5. Shell thickness after improved process

4. conclusion

In view of the problem that the initial construction process is greatly affected by human factors, the single-sided vacuum adsorption molding method is adopted by drawing on the advanced achievements at home and abroad, the thickness of the fabricated FRP shell is uniform within 3mm, the sound transmission coefficient reaches 0.98-0.99, High sound transmission and comprehensive index requirements for anti current and vibration damping performance are realized.

References

1. Difficult Analysis of FRP Molding Chemical Industry Press, 2016.
2. Raw Materials and Selection of FRP. Chemical Industry press, 2009.

3. QC achievement of pipe curtain jacking accuracy in soft soil area. Shanghai Construction Science and Technology. 2022 (01).
4. Design of a new sealing structure of glass fiber reinforced plastic dome and its model test. Journal of Naval University of Engineering. 2004(04).
5. Improvement of surface coating effect of special vehicle electric vehicle FRP shell. Modern paint and coating. 2018, 21(03).

Research on seismic behavior of prefabricated building joints in China

Hejun Zhang*

School of Civil Engineering, Architecture and Environment, Hubei University of Technology, Wuhan, China

Abstract. The rapid development of China's construction industrialization process has led to increased research into prefabricated building structures. In comparison to cast-in-place construction, prefabricated buildings offer many advantages. Increasingly important has been research on prefabricated building joints' seismic performance. A prefabricated beam, slab, column, and wall's connection affects its seismic performance, and each joint has different characteristics. In light of this, this paper analyzed the seismic performance of prefabricated joints based on their development and application.

Keywords: Seismic performance, Prefabricated building, Construction joints.

1. Introduction

China has been using cast-in-place concrete systems for its building structures ever since the 1980s, and most of the buildings are constructed using these systems [1]. Since the majority of the concrete casting work was done manually, the process of casting concrete on site was noisy and dusty. It also needed a number of workers, had long working cycles, was very intensive in terms of construction, and consumed a great deal of materials [2]. With the acceleration of China's modernization process and the improvement of production technology, prefabricated building structures have become a popular option in the engineering industry due to their short construction duration, quick return on investment [3], and simple and fast construction methods, as well as their high efficiency in utilizing building materials. In spite of this, there are certain problems that are associated with the use of prefabricated building components in the engineering field. Uneven seismic performance of building components is one of the most troublesome problems that can occur [4]. As a result of this, the development of prefabricated building components has been greatly restricted. The study of the seismic performance of joints has now become the key to advance prefabricated building components to the next level of development.

Currently, prefabricated buildings are widely used across the world. This paper examines a few common types of joints that are commonly used in prefabricated buildings in China. There will also be an analysis of the seismic behavior of each type of joints. The purpose of this paper is to argue that the most commonly used joints in China are welding joints, bolt joints, cross-lap joints and ring-stiffened joints, each of which has its own advantages and disadvantages, yet each type is suitable for a different application.

2. Analysis on Different Joints of Prefabricated Buildings

There are two main types of joints in prefabricated buildings: structural joints and non-structural joints. A structural joint refers to a connection between a column and a column, a column and a beam, and a beam and a secondary beam, while a non-structural joint refers to a connection between an external slab (wall). A house's connection between prefabricated columns and columns is one of its most important defenses against earthquakes, hurricanes, and other natural disasters. In addition, prefabricated beams and columns are adjacent, and are connected by prefabricated beam-column joints. As a result, the joints of prefabricated beams and columns play an irreplaceable role in the overall structural load-bearing performance, structural stiffness, and seismic behavior of buildings [5]. A common building design in China utilizes cast-in and shelf connections to connect the primary and secondary beams, whereas the reasonable design of the prefabricated joints is also crucial to the overall seismic performance of the structure. According to the form of external hanging, non-structural joints are usually constructed using external hanging plates or external lifting, which can broadly be categorized into point hanging, line hanging, and the combination of point and line. In non-structural joints, prefabricated elements are often used to join wall panels with concrete frame structures, and good prefabricated joints can greatly improve the house's seismic performance.

* Corresponding author: 2430360323@qq.com

At present, the joints of prefabricated buildings in China can broadly be classified into two categories, namely the connections between the frame beams and columns and the connections between the wall panels of the building. According to different construction methods, the connection of frame beam-column nodes can be classified into two kinds: dry joints and wet joints. Accordingly, the dry joints are used for the connection of horizontal joints and vertical joints between prefabricated assembled components and units, and the wet joints are used for the connection of horizontal joints.

2.1 Dry Joints

A dry joint is basically used to maintain and control the connection between two prefabricated elements by mechanically connecting the reinforcement bars of the two prefabricated elements to each other [6]. A mechanical foundation consists of many structural elements and elements interconnected by joints of this type. There are several common methods of dry jointing that are used, including mechanical sleeve joints, prestressing crimp joints, bull-leg joints, welded joints, and bolted joints. In the case of dry joints, they are fast and easy to construct as well as having excellent stiffness and load-bearing properties, which are in line with the expectations of modern housing construction in China. However, their ductility is poor and their recovery force is insufficient, so they have a certain gap in comparison to cast-in-place prefabricated frame structures and are more difficult to protect against damage caused by earthquakes and other forces.

In the past, dry joints have been used with the bridge segments being formed by match casting, which has been an effective way of creating bridge segments [7]. It is thought that the reason for the prevalence of the technique in the past was its lower cost and shorter construction time. This joint does not have any gluing material to keep it sealed up. This means that leakage through the joint into the box culvert may occur from time to time, which, in turn, may have a detrimental effect on the durability of the external post-tensioning tendons. Furthermore, it has been found that under certain conditions, such as seismic, temperature, and creep, the joints tend to open because of these effects. Additionally, there were reports of concrete spalling at the intersection of the bridge and the top slab.

2.2 Wet Joints

The wet joint consists of two precast segments that are glued together with epoxy glue at the joint points. During the process of installing epoxy glue, a temporary precompression pressure of 0.3MPa will be applied at the top and bottom of the mating precast segments by means of stress bars at the top, bottom, and sides of the segments. Under the application of pressure, the epoxy sets and becomes solid. In the use of epoxy joints, the lubrication helps in the fit-up and alignment of the mating segments between the segments, and minimizes the impact of hard points on the mating segments as a result of the contact between the segments.

3. Seismic Behavior Analysis

The degree of seismic behavior of prefabricated building components depends in a great deal on the reasonableness of the design and the ease of construction of the joints of the prefabricated building components. This section aims to investigate the seismic performance of prefabricated buildings by analyzing several typical connections methods of prefabricated building components through the analysis of the seismic performance of joints in prefabricated buildings.

3.1 Welding Joints

While the welding joints of prefabricated components have the same stiffness and strength as those of cast-in-place joints, their load bearing capacity depends to a large extent on the strength and rigidity of the side plates. It may lead to deformation of the welding joints if the side plate is not installed in a reasonable manner. In addition to that, welding joints are becoming more demanding for the welding process, and the welded points may cause cracks and damage to the system as a result [8]. Whenever a building is deformed under the influence of earthquakes, it will become more prone to partial cracks and residual stresses, which will have a significant impact on the building's seismic performance. A number of surveys have indicated that welding joints exhibit undesirable seismic behavior and are susceptible to brittle damage.

3.2 Bolt Joints

It is often the case when bolt joints are used in the construction of prefabricated components that, despite their complex structures, their installation is relatively straightforward. In order to better ensure that prefabricated structures that are needed to withstand dynamic forces are able to do so, bolt joints are often used to construct these structures because of their relatively high capacity for handling such forces. A number of researchers have been investigating the combination of welding and bolt joints in recent years in order to improve the seismic behavior of structures [9].

3.3 Cross-lap Joints

It is a common practice in real engineering practice to use cross-lap joints. Using this method, the stronger the concrete is in terms of its structural performance, the better the seismic performance of the specimen will be [10]. Moreover, the piercing steel plate has a significant impact on the structure of the building as well. As the thickness of the piercing steel plate increases, so will the seismic capacity and load-bearing capacity of the prefabricated frame components which are fabricated from this material. In the construction of prefabricated concrete buildings, this type of construction is now widely used to connect the joints of the buildings.

3.4 Ring-stiffened Joints

Due to the accelerated pace of modernization, ring-stiffened joints are increasingly being used as a means of assembling prefabricated components together. The stiffened rings' performance is largely dictated by their stiffness, strength, load bearing capacity, ductility, and other characteristics, which all contribute to their effectiveness. In spite of their robustness to seismic forces, ring-stiffened joints still exhibit significant pinching as the load increases, despite their good resistance to seismic forces. In turn, this leads to improved seismic performance of the joints as a result. This type of joint, however, requires a high level of processing technology and a relatively high cost in terms of its production.

4. Conclusions

Prefabricated building joints should be designed in accordance with the actual structure and conditions of the building. This includes choosing the correct prefabricated connector method and structural elements according to the relevant seismic level. While designing the joints, it is essential to consider the relevant practical scenarios to ensure that they have sufficient stiffness, strength, bearing capacity, ductility, and overall stability. It is also important to incorporate the seismic performance improvement into the conceptual design process. Prefabricated buildings must be engineered to have a high degree of seismic resistance and low circumferent loads. This is done by increasing the seismic performance of the components while improving the overall seismic resistance of the building.

References

1. Zhao, G., Yang, Y., Zhang, H., & Zhang, G. (2019). A case study integrating field measurements and numerical analysis of high-fill slope stabilized with cast-in-place piles in yunnan, china. Engineering Geology, 253, 160-170. https://doi.org/10.1016/j.enggeo.2019.03.005.
2. Yang, J., Liang, S., Zhu, X., Sun, C., & Guo, Z. (2017). Seismic behavior of precast concrete coupled shear walls with engineered cementitious composite (ECC) in the critical cast-in-place regions. Science China. Technological Sciences, 60(8), 1244-1254. https://doi.org/10.1007/s11431-016-9052-2.
3. Lin, X., & Jie, S. (2019). Analysis of the development and trend of china's prefabricated buildings. IOP Conference Series. Earth and Environmental Science, 371(2), 22043. https://doi.org/10.1088/1755-1315/371/2/022043.
4. Furnish et al. American Institute of Physics, Melville, NY, 2002, pp. 651–654. Filiatrault, A., & Sullivan, T. (2014). Performance-based seismic design of nonstructural building components:The next frontier of earthquake engineering. Earthquake Engineering and Engineering Vibration, 13(S1), 17-46. https://doi.org/10.1007/s11803-014-0238-9.
5. Wang, Q., Senatore, G., Jansen, K. M. B., Habraken, A., & Teuffel, P. (2021). Seismic control performance of a three-story frame prototype equipped with semi-active variable stiffness and damping structural joints. Earthquake Engineering & Structural Dynamics, 50(13), 3379-3402. https://doi.org/10.1002/eqe.3514.
6. Oettel, V., & Empelmann, M. (2019). Structural behavior of profiled dry joints between precast ultra-high performance fiber reinforced concrete elements. Structural Concrete : Journal of the FIB, 20(1), 446-454. https://doi.org/10.1002/suco.201800117.
7. Foxell, S. (2018). Professionalism for the built environment. CRC Press. https://doi.org/10.1201/9781315707402.
8. Zhang, L., Li, X., Nie, Z., Huang, H., & Niu, L. (2016). Comparison of microstructure and mechanical properties of TIG and laser welding joints of a new Al–Zn–Mg–Cu alloy. Materials & Design, 92, 880-887. https://doi.org/10.1016/j.matdes.2015.12.117.
9. Liu, F., Shan, M., Zhao, L., & Zhang, J. (2018). Probabilistic bolt load distribution analysis of composite single-lap multi-bolt joints considering random bolt-hole clearances and tightening torques. Composite Structures, 194, 12-20. https://doi.org/10.1016/j.compstruct.2018.03.100.
10. Okawachi, T., Ishihata, K., Nomoto, N., Tezuka, M., Kamikuri, Y., Nozoe, E., & Nakamura, N. (2020). Using three-dimensional nasal forms to compare definitive unilateral cleft lip nose correction with/without a cross-lap joint cartilage graft technique. Journal of Cranio-Maxillo-Facial Surgery, 48(11), 1035-1044. https://doi.org/10.1016/j.jcms.2020.08.008.

https://doi.org/10.1051/matecconf/202338001005

Simultaneous Determination of nicotinamide, kojic acid, Tranexamic acid, raspberry glycoside, azelaic acid, magnesium ascorbate phosphate and β- Arbutin in whitening cosmetics by UPLC-MS/MS

Xiao Zou [1,2], Lixiu Yan [1], Xiaofang Luo [1], Enting Wang [1], Mingjuan Zhang [1,2], Wuping Huang [1], Wulin Li [3], Lanyan Tong [1,2]

[1] Chongqing Academy of Metrology and Quality Inspection, Chongqing, 401123, China;
[2] Chongqing Engineering Research Center for Food Safety, Chongqing, 401123, China;
[3] Chongqing Jinzhi Quality Certification Limited Company, Chongqing, 401122, China

Abstract. A UPLC-MS/MS method was developed for simultaneous determination of 7 whitening ingredients: nicotinamide, kojic acid, Tranexamic acid, raspberry glycoside, azelaic acid, magnesium ascorbate phosphate and β-Arbutin in cosmetics. The whitening active components were extracted from cosmetics by supersonic extraction with sodium chloride and dichloromethane to disperse the sample, and supersonic extraction with 0.015 mol/L potassium dihydrogen phosphate solution purified by HLB solid phase extraction column, scanned and detected by electrospray ionization source with positive and negative ion alternate scanning mode and multiple reaction monitoring mode. The results showed that the whitening active ingredients were separated within 3 minutes, with a good linear relationship (R>0.999), and the detection limit was 0.10mg/kg~0.75mg/kg. The recoveries (n=6) were 78.84%-104.85%, and the RSDs were 0.24%-11.35%. This method is suitable for the rapid determination of whitening active ingredients in cosmetics.

Key words: Whitening active; Cosmetics; UPLC-MS/MS; Rapid Determination.

Whitening and freckle-removing cosmetics have the effect of preventing spots and pigmentation caused by sunlight or other reasons, so they are favored by consumers. Its mechanism of action is to interfere or even block the formation of melanin. According to the principle of action of different chemicals, the active ingredients of whitening and freckle-removing can be classified into three categories [1]: one is as reducing agent, including vitamin C, E and derivatives; The second category is enzyme inhibitors: mainly inhibiting tyrosinase activity and affecting the transport of melanin, mainly represented by hydroquinone, azelaic acid, arbutin, kojic acid and derivatives; The third category is stripping agent: linoleic acid, citric acid, placenta extract, etc. Nicotinamide has been widely used in the medical and beauty fields with its anti-aging effect. It is believed that as a B-group vitamin compound, nicotinamide can achieve the effect of whitening skin by inhibiting the formation and transfer of melanin particles and accelerating the renewal of skin cells [2]. The cross and excessive use of these new whitening active ingredients will interfere with the normal enzyme conversion of tyrosine to melanin in the skin, cause adverse reactions to the skin, and even cause cancer [3].

At present, the commonly used methods for detecting whitening active ingredients in cosmetics include thin layer chromatography (TLC), gas chromatography (GC), high performance liquid chromatography (HPLC), gas chromatography-tandem mass spectrometry (GC-MS/MS), liquid chromatography-tandem mass spectrometry (HPLC-MS/MS) [4-5], etc. Cosmetics are difficult to separate and purify due to the variety of raw materials and ingredients, complex matrix, and the removal and separation of impurities and interfering substances in the pretreatment will affect the accuracy of the results. Ultra-high performance liquid chromatography-mass spectrometry/mass spectrometry has the advantages of high efficiency, fast, multi-component analysis, high sensitivity, and accurate quantification. At present, there are few studies on the analysis and detection of whitening ingredients. In this study, three commonly used matrices in whitening cosmetics: aqueous solution, lotion, and face cream were investigated, and the chromatographic conditions for the separation of seven whitening active ingredients were optimized.

1. Experiment

1.1 Instruments and reagents

Electronic balance (MS105, Mettler); Liquid mass spectrometry (AB4500, Agilent 1290); Ultra-pure water machine (PC1ANRXM2 (Chorus 1), ELGA, UK); Ultrasonic cleaner (KQ3200DE, Kunshan Ultrasonic Instrument Co., Ltd.); Vortex mixer (IKA VORTEX3).
Methanol (HPLC grade, Norsch), acetonitrile (HPLC grade, Shanghai Xingke High Purity Reagent Co., Ltd.), formic acid (HPLC grade, Shanghai Ampu), acetic acid (HPLC grade, Shanghai Ampu), dichloromethane (HPLC grade, Norsch), ammonium acetate HPLC grade, Shanghai Ampu), oxalic acid, sodium chloride, potassium dihydrogen sulfate, nicotinamide (NA, purity ≥ 98%), kojic acid (KA, purity ≥ 98%) Tranexamic acid (TAA, purity ≥ 99%), raspberry glycoside (RG, purity ≥ 98%), azelaic acid (AA, purity ≥ 98%), magnesium ascorbate phosphate (MAP, purity ≥ 98%) β- Arbutin (AB, purity ≥ 98%).
The cosmetics samples are all from the products sold in Chongqing. The mall randomly selects and purchases the water, lotion and face cream products with the words "whitening" on the label.

1.2 Preparation of standard solution

Accurately weigh NA, KA,TAA,RG, AA, MAP and AB standards 10mg (accurate to 0.1mg) is put into different 10mL brown volumetric flasks, and the chromatography-pure methanol is dissolved to volume, and the concentration is 1000 μg/mL of single standard stock solution, stored under 4 °C in dark. Accurately transfer 0.05mL of single standard stock solution into the same 50mL volumetric flask, and obtain the concentration of 1 after the volume of methanol is fixed μ G/mL mixed standard intermediate solution, stored under 4 °C in dark. Take an appropriate amount of mixed standard intermediate solution, and prepare 0.1% oxalic acid solution into mixed standard working solution of appropriate concentration, which will be used on the machine and ready for use.

1.3 Sample treatment

1.3.1 Extraction

Accurately weigh 1.000g of sample into a 50mL centrifuge tube with a stopper, add 2 g of sodium chloride and 5 mL of dichloromethane, place a ceramic homoproton, shake it with a vortex mixer until the sample is dispersed, and then add 20 mL of 0.015mol/L KH2PO4 solution for ultrasonic extraction for 30 minutes, and centrifuge at 8000 r/min for 3 minutes, transfer the upper supernatant into a 50mL volumetric flask, and extract the lower substances once with 20 mL of KH2PO4 solution, After the same centrifugation operation, combine the supernatant into the same volumetric flask, fix the volume to the scale, and wait for purification.

1.3.2 Purification

Transfer 4mL of the above extraction solution to the activated HLB column (200mg/6mL, activated successively with 3mL methanol and 3mL ultra-pure water in advance), control the flow rate to 1mL/min, collect the effluent, elute with 10mL 2% acetic acid-methanol solution, collect the eluate at 45 °C, evaporate and concentrate it to near dry, dissolve the residue with 1.0mL 0.2% formic acid-acetonitrile solution, and pass 0.22 μ M filter membrane, UPLC-MS/MS analysis.

1.4 Instrument conditions

1.4.1 UPLC conditions

Chromatographic column: C18, 50mm × 2.1 mm ×1.9 μm; The column temperature is 30 °C; Injection volume 10 μL; The mobile phase A is acetonitrile, and the mobile phase B is 5mmol ammonium acetate (containing 0.2% formic acid); Flow rate: 0.30mL/min; The gradient elution procedure is shown in Table 1.

Tab. 1 Elution program of 7 kinds of whitening active components

Time (min)	Acetonitrile (%)	5 mmol ammonium acetate (containing 0.2 % Formic acid) (%)
0	40	60
2.7	80	20
5.0	100	0
6.5	40	60

1.4.2 Mass spectrum conditions

Electric spray ion source (ESI); Simultaneous scanning of positive and negative ions; Monitoring mode: multiple response monitoring (MRM); Ion source temperature (TEM): 450 °C; Curtain Air (CUR): 35 Psi; Collision gas (CAD): 9mL/min; Electric spray voltage (IS): - 4500V and 5500V; Spray gas (Gas1): 50Psi; Auxiliary heating gas (Gas2): 50Psi.

2. Results and discussion

2.1 Optimize mass spectrum conditions

At the concentration of 1μg/mL, Seven kinds of mixed standard intermediate solution samples were scanned under different conditions of positive and negative ions, and the ionization mode was determined by the response strength of the ions. At the same time, the mass number of the precursor ion-product ion was optimized and screened successively in a triple four-stage bar to determine the quantitative and qualitative ion pairs, and the collision energy (CE) and retention time were optimized. Finally, the optimized mass spectrum conditions of the seven whitening active ingredients were obtained, The results are shown in Table 2. The typical MRM chromatogram detected under the optimal mass spectrum conditions is shown in Figure 1.

https://doi.org/10.1051/mateconf/202338001005

Tab. 2 Optimized mass spectrometry conditions for 7 kinds of whitening active ingredients

components	retention time (min)	precursor ion (m/z)	product ion (m/z)	Dwell time (mSec)	Q1Pre (V)	CE (eV)	Q3Pre (V)	Scan mode
NA	0.75	123.00	80.05*	20.0	-10.0	-21.0	-14.0	Positive ion
		123.00	78.05	20.0	-10.0	-24.0	-13.0	
KA	0.69	143.00	69.05*	20.0	-12.0	-17.0	-12.0	Positive ion
		143.00	97.05	20.0	-30.0	-18.0	-19.0	
TAA	0.49	158.10	95.10*	20.0	-20.0	-16.0	-20.0	Positive ion
		158.10	67.05	20.0	-13.0	-27.0	-11.0	
RG	1.66	344.15	107.05*	20.0	-21.0	-21.0	-19.0	Positive ion
		344.15	165.10	20.0	-21.0	-12.0	-10.0	
AA	0.52	187.00	125.10*	20.0	14.0	16.0	13.0	anion
		187.00	169.10	20.0	14.0	15.0	11.0	
MAP	0.47	255.00	78.90*	20.0	18.0	31.0	17.0	anion
		255.00	236.90	20.0	10.0	15.0	27.0	
AB	0.67	271.10	108.05*	20.0	20.0	26.0	20.0	anion
		271.10	161.00	20.0	13.0	8.0	17.0	

Note: The ions with * are quantitative ions

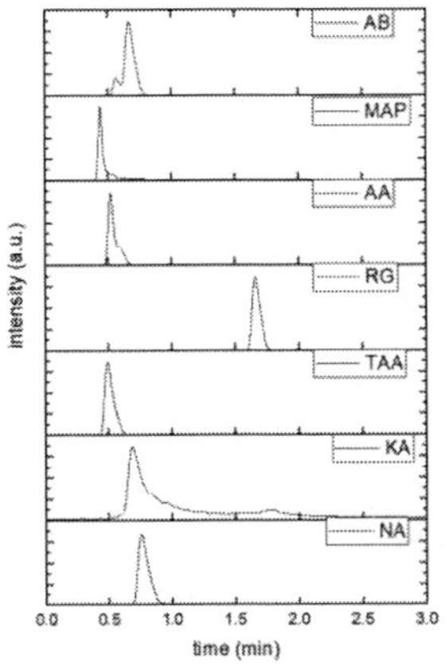

Fig. 1 MRM of 7 kinds of whitening active components

2.2 Linear relationship and detection limit

Use the mixed standard working solution series of different concentrations to inject samples under the optimal conditions to obtain the standard spectrum of whitening active ingredients of different concentrations. Take the concentration of each compound as the abscissa and the peak area of its corresponding quantitative ion pair as the ordinate to draw the standard curve, and obtain the linear range, regression equation and correlation coefficient R of seven whitening active ingredients, The detection limit LOD of 7 whitening active ingredients is calculated from the signal to noise ratio of their chromatograms greater than 3. Results All substances had a good linear relationship (R>0.999), and the detection limit was 0.10 mg/kg~0.75 mg/kg, as shown in Table 3.

Tab. 3 Linearity and detection limit of 7 kinds of whitening active components

components	Linear regression equation	correlation coefficient(R)	linear range(mg/L)	LOD/(mg/kg)
NA	y =2851240x+95401.2	0.9996	0.01~0.5	0.10
KA	y =142532x+2105.4	0.9995	0.02~1.0	0.35
TAA	y=4172300x+100119.4	0.9997	0.01~0.5	0.10
RG	y = 235852x+1501.9	0.9996	0.02~1.0	0.35
AA	y = 25575x+1523.1	0.9996	0.02~1.0	0.35
MAP	y = 7623x+242.1	0.9993	0.05~2.5	0.75
AB	y = 28180x+213.45	0.9995	0.02~1.0	0.35

2.3 Recovery and precision

Under the optimal conditions, select three kinds of negative samples, weigh appropriate amount of samples respectively, add mixed standard solution with three concentration gradients of high, middle and low, and carry out six parallel tests at each gradient level, and calculate the recovery rate and precision. The results showed that The recovery rates of nicotinamide, kojic acid, tranexamic acid, raspberry glycoside, azelaic acid, magnesium ascorbate phosphate and β-arbutin in water, lotion and face cream are 78.84%~104.85%, and the relative standard deviations (RSD) are 0.24%~11.35%, all less than 15%.

Tab. 4 Recovery and precision of 7 whitening active components (n=6)

components	Add level (mg/kg)	solutions		lotion		face cream	
		recovery rate (%)	RSD (%)	recovery rate (%)	RSD (%)	recovery rate (%)	RSD (%)
NA	0.5	94.22	2.42	93.48	3.52	95.24	4.56
	1.0	95.28	2.01	90.53	3.52	93.14	3.25
	5.0	97.85	1.59	93.66	0.24	100.01	2.48
KA	0.5	90.54	4.38	88.54	5.24	85.49	4.12
	1.0	89.64	3.98	86.28	2.65	90.54	3.58
	5.0	99.47	2.83	98.86	2.30	104.24	1.05
TAA	0.5	90.36	3.42	88.67	6.87	92.54	4.24
	1.0	92.15	2.85	86.01	3.54	95.16	3.56
	5.0	104.26	1.68	98.53	3.24	102.46	0.89
RG	0.5	85.64	8.42	80.67	6.57	83.26	6.36
	1.0	88.39	5.24	85.26	5.42	88.96	7.59
	5.0	92.44	2.62	92.22	2.68	96.24	4.51
AA	0.5	82.68	5.96	85.46	5.65	83.26	5.24
	1.0	88.54	3.24	88.96	3.25	90.26	3.26
	5.0	92.46	1.26	96.16	2.15	90.56	1.26
MAP	1.0	78.84	11.35	79.52	9.46	80.14	9.98
	2.0	82.12	8.67	80.46	6.41	83.20	7.85
	10.0	85.23	6.42	82.49	4.21	84.06	3.59
AB	0.5	83.26	5.61	88.57	8.59	84.68	6.05
	1.0	94.26	4.24	84.26	5.49	93.01	4.18
	5.0	104.85	1.16	93.26	1.08	98.71	3.06

2.4 Contents of Whitening Ingredients in Commercial Products

Nine kinds of aqueous solutions, lotion and face cream with the words "whitening" on their labels purchased in major markets of Chongqing were determined. The results showed that among the nine samples, ascorbic acid phosphate salt was detected in No. 2 aqueous solution, but it was less than the detection limit, and could not be quantified; No. 5 lotion detected 20 mg/kg of tranexamic acid and 7.8 mg/kg of ursolic acid; Nicotinamide 25 mg/kg and azelaic acid 19 mg/kg were detected in No. 8 face cream.

3. Conclusion

In this paper, a method for simultaneous determination of seven whitening active ingredients in cosmetics by UPLC-MS/MS was established. The whole experiment is simple and fast, with high accuracy and sensitivity. It is suitable for the detection of multi-component whitening active ingredients in cosmetics with whitening and freckle removing effects, such as water, lotion, face cream, etc. It can achieve rapid qualitative and quantitative detection within 3 minutes, greatly shorten the detection time, improve the detection efficiency, and verify the linear equation The detection limit, recovery rate and precision indicators can meet the needs of daily market supervision, and provide technical support for preventing the risk of abuse of chemical substances and improving the supervision ability.

Acknowledgements

Fund projects: Project of Scientific Research Program of Chongqing Municipal Bureau of Quality and technology supervision (CQZJKY2018009).

References

1. Couteau C, Coiffard L. Overview of skin whitening agents: Drugs and cosmetic products[J]. Cosmetics, 2016, 3(3): 27.
2. Nordin F N M, Aziz A, Zakaria Z, et al. A systematic review on the skin whitening products and their ingredients for safety, health risk, and the halal status[J]. Journal of cosmetic dermatology, 2021, 20(4): 1050-1060.
3. Burger P, Landreau A, Azoulay S, et al. Skin whitening cosmetics: Feedback and challenges in the development of natural skin lighteners [J]. Cosmetics, 2016, 3(4): 36.
4. Liu Y, Liu D, Xie Z. Rapid and Specific Fluorescence Method for the Quantification of Arbutin in Cosmetics[J]. Analytical Letters, 2022, 55(2): 318-326.
5. Li Shasha, Yin Bo, Zhai Honglin, et al. An effective approach to the quantitative analysis of skin-whitening agents in cosmetics with different substrates based on conventional UV-Vis determination[J]. Analytical Methods, 2019, 11(11): 1500-1507.

Research and Development of Intelligent Drying Detection Technology for Machine-Made Sand

Shaoying Jia*, Xiong Zhang and Yelin Lu

Huahao Zhongneng Testing Technology Group Co., Ltd. Beijing 100086, China

Abstract. This paper aims to develop a high precision variable dielectric constant capacitor detector for mechanical sand water content testing. The non-contact on-line test method can effectively solve the problems of low repeatability of test data, large hysteresis error and poor linearity due to the non-uniform water content. In terms of hardware, CAV444 is a high-precision integrated chip, which matches the corresponding test circuit. The system communicates with PC through serial interface and implements the man-machine interface of PC system based on VS. Secondly, the distribution map method is used to measure the water content of machine-made sand, and the error analysis is carried out. The test shows that the output of the capacitive hygrometer is 0-5 V. The delay time should not exceed 0.15%. The repetition rate was within 0.33%. The measurement accuracy of this method is less than 3.52%.

Keywords: mechanical sand; water content; Intelligent monitoring; on line monitoring; capacitor type sensing.

1. Introduction

In cement production, the measurement of water content of mechanical sand is the key factor affecting the final quality of concrete. Using the traditional drying method to determine the moisture content is not only time-consuming but also easy to cause the sample is not typical. This method can realize the real-time monitoring of moisture content with higher frequency, better representativeness, higher real-time performance and lower error. In cement production, an important technical development is to change the traditional drying method to on-line moisture measurement, in order to improve the real-time moisture measurement and improve the accuracy of match comparison. Although there are special workers to test the moisture content of fine aggregate powder in the production of concrete mixing field, it can only be carried out twice a day. In order to obtain the data faster, the fine aggregate is usually heated by microwave, which will cause the complicated process, resulting in artificial errors, thus affecting the water content of the product. In the laboratory, the oven is usually used for determination, also known as the standard oven drying, is currently a measurement method widely used in the world. In this case, real-time monitoring technology comes into being [1]. The method of measuring moisture and moisture had been used since the 1940s, beginning with the measurement of moisture in soil and paper. At the beginning, the testing equipment was very simple, consisting only of an antenna and a waveguide containing samples, and the accuracy of the testing equipment was very poor. Therefore, before this, Online measurement of water content has not been well developed. Due to the progress of technology and the development of technology, the determination of moisture content has been widely used, such as crude oil, seeds, tobacco, etc., the application of circuit is more and more widely, MOSFET and IC and other modern components come out, so that the test method has been greatly developed. The basic principle of microwave testing has also been developed from the original resonant cavity method to microstrip technology, expanding from 300 KHz to 20 GHz, and the test field has changed from the original test without EMI to the harsh production conditions. In this paper, based on the capacitor, the use of monolithic microprocessor technology, developed a mechanical sand moisture detection device suitable for large-scale engineering construction site, real-time detection from the source, wear small, high precision, can effectively reduce the cost, improve production efficiency, for enterprises to obtain good economic and social benefits.

2. Design of intelligent water content measurement system for machine-made sand

2.1 Circuit Design

Microwave mixing technology and IQ mixer (AD8349) are used to mix the basic frequency of 400 KHz and the carrier signal of 2.4 GHz and transmit and modulate in the

* Corresponding author: 34080575@qq.com

https://doi.org/10.1051/matecconf/202338001006

transmission array. Demodulation is performed by IQ demodulator (AD8347) and compared with oscilloscope. Figure 1 shows the high frequency circuit of the microwave.

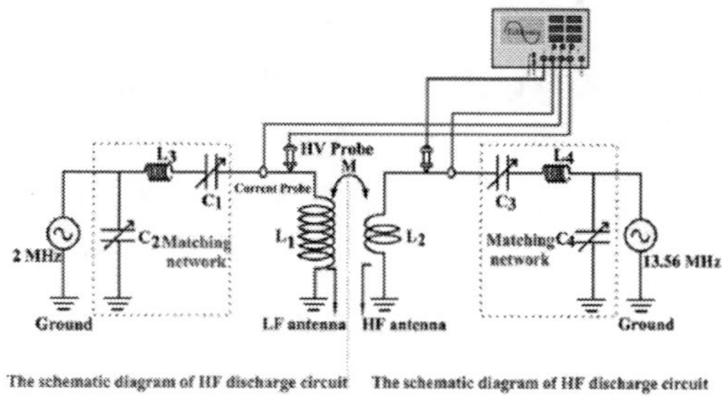

Figure 1. Schematic diagram of high frequency circuit

The microwave circuit's electronics include ADF4351, filter, isolator, power divider, amplifier, Q demodulator, IQ mixer, and so on. ADF4351 is a kind of integrated VCO with external filtering and baseband frequency and has the effect of noise reduction. It can be operated by pins and software and has an auxiliary transmitting power source [2]. The filter used in the test is a band-pass filter, only 2.45 GHz+500 MHZ microwave signal can pass through, other interference is eliminated. The experiment uses a 14 dBi antenna, which is characterized by a single direction of radiation and electromagnetic wave receiving especially strong, its transverse and longitudinal angles are 30°; IQ demodulators are mainly used to modulate and filter mixed signals to get a weakened baseband. The IQ mixer is used to mix the base frequency of 400 KHz at 2.4 GHz with the carrier signal at 2.4 GHz. Here, the IQ mixer and demodulation machine works as shown in Figure 2.

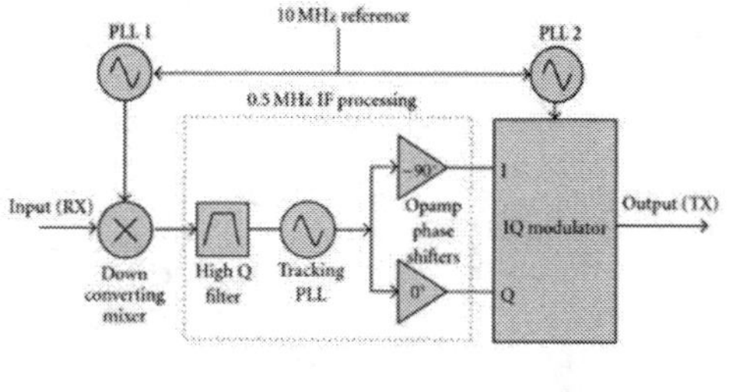

Figure 2. Schematic diagram of IQ mixer

The incoming signals are two orthogonal signals. They're called $B\cos\omega_B t$ and $B\sin\omega_B t$. Carrier signals are $A\cos\omega_A t$ and $A\sin\omega_A t$, and Y signal is obtained after the signals are multiplied and added by the multiplier and adder respectively. Its expression is shown in Equation (1)

$$Y=\frac{1}{2}\left(B\cos\omega_B t\bullet A\cos\omega_C t-B\sin\omega_B tA\sin\omega_C t\right)$$
$$=\frac{AB}{2}\cos\left(\omega_B+\omega_C\right)t \qquad (1)$$

The measurement of phase shift is the key to determine the water content of fine bone. In the experiment, the thickness of machine-made sand should be less than the wavelength of microwave as far as possible [3]. Due to the short wavelength and poor penetration of microwave, too thick sample will lead to microwave signal reflection, affecting the experimental results. In addition, certain phase compensation is required for the calculated phase shift results, as shown in Equation (2).

$$\Delta\phi=\phi_s-\phi_0-2\pi n \qquad (2)$$

2.2 Working principle of the sensor

The purpose of this paper is to detect the change of capacitor capacity due to the difference of sand moisture content by using the boundary action between the two ends of the capacitive sensor, so as to reduce the sand wear. In addition, it is also necessary to develop a testing mechanism, which can install the capacitive sensor based on the edge action principle into the feed hole and detect the water content of the feed sand in real time. The CAV444 chip is used to collect the capacitor's electric quantity, and then the obtained electric quantity is converted into a voltage signal that can be read through the signal conversion circuit [4]. Then STM32 is used as the main control unit. By communicating with PC and displaying it on PC, real-time and real-time detection of flow can be realized. This capacitive sensor has excellent accuracy and is harmless to human body. It can measure the moisture content of sand grains in real time to improve the quality of cement after mixing.

3. System architecture and function realization

3.1 System Architecture

This project makes use of the edge effect between the two plates of the capacitor sensor, so that the sand can measure the change of the capacitance value caused by the different water content of the sand without touching the plate, and effectively reduce the wear. At the same time, it is fixed at the feed port to realize real-time measurement of the water content of the feed sand (Figure 3). The variation of capacitance value detected by the measuring device designed based on capacitance edge effect can be changed by using sand with different water content [5]. The capacitance is collected by CAV444 chip, and then converted into readable voltage signal output through the supporting signal conversion circuit. After a series of circuit design, calibration and fitting including STM32 main control chip circuit, general voltage regulator power circuit, USB program download circuit, clock reset circuit, etc., the serial port is directly communicated with the PC end, and displayed on the display screen. Finally, the online real-time measurement of the mechanical sand water content passing through the measuring plate is realized.

 https://doi.org/10.1051/matecconf/202338001006

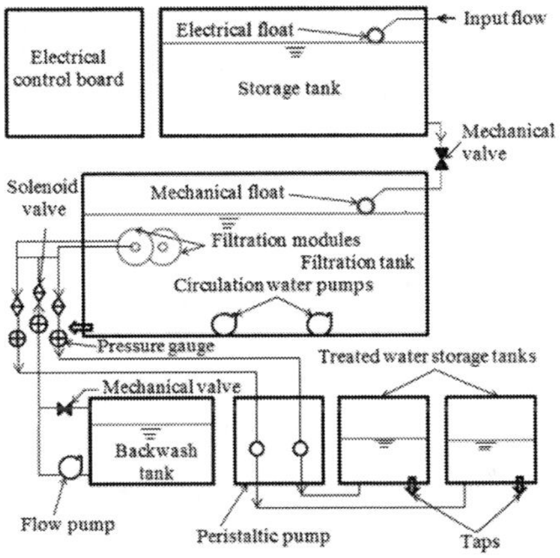

Figure 3. System block diagram

3.2 Design of data acquisition module

On this basis, the test results of the capacitive sensor are used to convert the capacitors of CAV444 into a voltage value, which is compared with the input reference voltage and then fine-tuned. The architecture of the entire data acquisition module is shown in Figure 4.

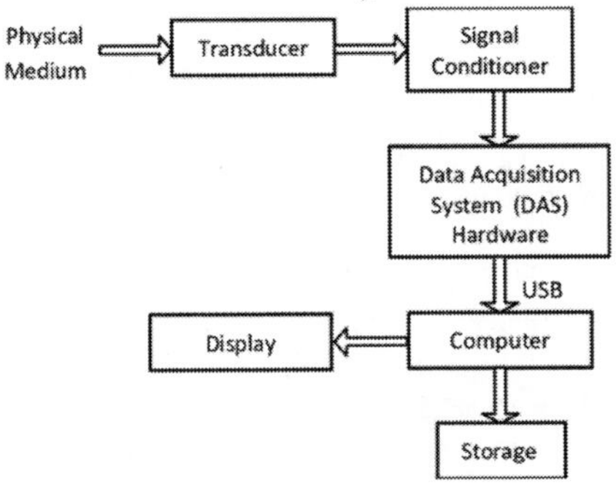

Figure 4. Structure diagram of data acquisition module

3.3 Design of data processing module

Due to its high sensitivity, high digital-analog transformation is required. The moisture content of mechanical sand is very low and difficult to determine accurately. The ADS1115 is a miniature 16-bit ADC with a built-in ADS1115 (ADC) converter chip that can transmit signals through CI2 compatible CI2 ports. Suitable for -40~-140 degrees Celsius [6]. The high measurement accuracy of 16 bits ensures the correctness of the system. The transistor pins of the ADS1115 are shown in Figure 5.

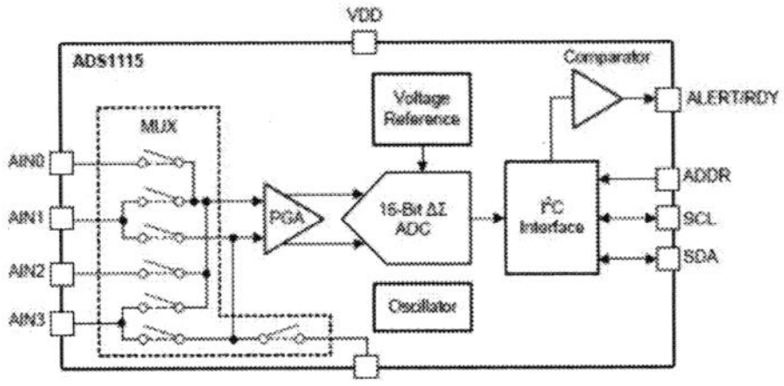

Figure 5. Pin diagram of ADS1115 chip

4. Test platform implementation

4.1 Static test platform

Figure 6 shows a microwave test bench. A sample box loaded with mechanical sand is placed in the middle of two antennas with the same orientation, and it is fixed on a horizontal aluminum plate with screws to ensure that the two antennas are on the same plane, so that the sending and receiving area of the two antennas is maximum, and the maximum microwave signal is obtained [7]. Aluminum alloy sheet is connected with the movable sliding device with screws, easy to adjust the test, so that it can move freely on the sliding track, so that the distance between it and the mechanical sand can be adjusted arbitrarily, to achieve the maximum frequency signal; After measuring the distance between the antenna and the mechanical sand, a locking mechanism is set on the sliding bar, so that the sliding block can be fixed at any time, so as to ensure the influence on the test piece and test piece during the test process, and reduce the deviation in the test process. The biggest advantage of the whole test platform is that it can be adjusted at will. When changing the antenna and adjusting the distance between the mechanical sand and the mechanical sand, the sliding device can be used for free movement, which greatly simplifies the debugging process of the test.

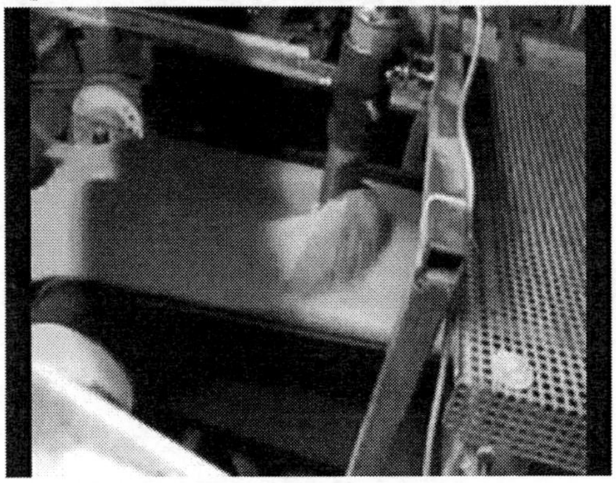

Figure 6. Microwave test platform

4.2 Dynamic test platform

In order to meet the requirements of industrialization, it is necessary to simulate the production situation of the enterprise and adopt dynamic method to test the determination of machine-made sand moisture. Mechanical sand is fed into the metering bucket through the silo. In order to monitor the water content of machine-made sand in real time, an antenna must be set at the outlet side [8]. In order to imitate the actual output, the materials in the silo are compressed according to a certain size. The overall height of the silo is 1800 mm, the width is 750 mm, the mouth of the silo is 1000 mm from the ground, the length of the discharging hole is 300 mm, the width is 200 mm, and the silo is loaded with 225 kg of machine-made sand. The mechanical sand was pushed down from the ground by means of hook traction. The overall silo is in the shape of a bucket, and mechanical sand is accumulated in the silo for a long time due to its large surface area, which is easy to adsorb on the side wall of the silo. It must be measured after mechanical sand filling (FIG. 7) [9].

Figure 7. Front view of dynamic test platform

5. System inspection

When correcting, there must be more than two centers. Each point is a non-converted reading taken by the material flow through the sensor as the actual moisture of the material after drying is obtained from the sensor. When using Hydro-Prob II for calibration, ensure that the materials tested and samples collected are representative. This is usually done using Hydronym's own Hydro-Com software and the calibrated data is stored in the sensor. As an example, data correction is performed by obtaining data from two points. As can be seen in FIG. 8, if the range of moisture content measured during calibration is relatively large, and there is a 0.3% deviation in the measured results in the high moisture area, the 0.3% deviation will decrease with the increase of moisture content (FIG. 8).

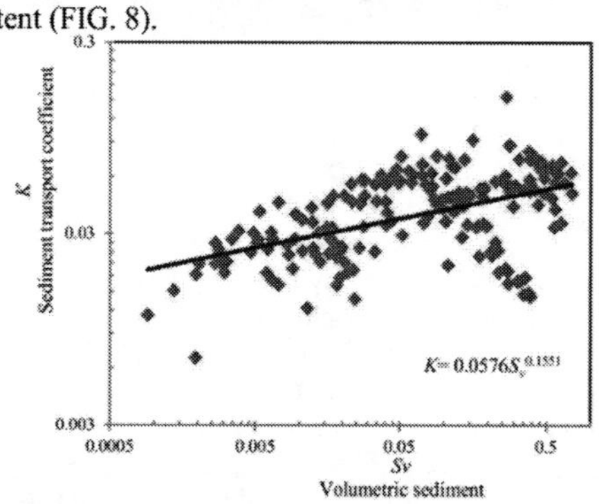

Figure 8. Schematic diagram of error control

6. Conclusion

This paper adopts a capacitive sensing probe with quadratic boundary effect, which can be used to detect the change of capacitance due to the difference of water content, so as to reduce the wear on the sand grains. By installing it in the feed port, the moisture content of the material can be detected in real time. Through different moisture content, the capacitance between the plates can be affected by different media.

References

1. Lin Jianhui, Wang Kai Chong. Effect of moisture content of machine-made sand on working performance and strength of wet mixed mortar. Guangdong Building Materials, vol. 38, pp. 37-44,October 2022.
2. Guo Yuanqiang. Study on the influence of water content change on the loose packing density of machine-made sand. Fujian Building Materials, vol. 7, pp. 26-27,January 2021.
3. Zhi Heng. Application of mechanized sand in concrete. Commodity and Quality · Construction and Development, vol. 11, pp. 7-8,March 2020.
4. Ma Haiyong. Study on the application of mechanized sand in filling of expressway subgrade. Building Construction, vol. 42, pp. 39-44,February 2020.
5. Zhong Guocai, Fan Zhi, Li Wenke, et al. Test and analysis of morphological characteristics of machine-made sand. Journal of Guizhou University: Natural Science Edition, vol. 37, pp. 79-84,January 2020.
6. Wang Zhen, Han Zili, Zhao Youming, et al. Study on particle morphology parameters of mechanized sand based on image analysis method. Railway Construction, vol. 60, pp. 59-64,November 2020.
7. Qi Sichao. Application of mechanical sand in PC concrete. Ju She, vol. 6, pp. 31-40,July 2020.
8. Peng Renchen. Analysis on the influence of stone powder content in mechanical sand on concrete properties. Fujian Building Materials, vol. 9, pp. 22-24,January 2020.
9. Liu Zhuang-jun. Study on the influence of curing method on the strength of mechanical sand concrete. Architecture · Building Materials Decoration, vol. 14, pp. 139-140, July 2020.

Numerical Analysis on Shock of Double-layer Plate with Foamed Aluminum

Peng Wang[1], Wei Huang[2], Zhao Li[1,*], Hao Zhang[1]

[1] Wuhan Second Ship Design and Research Institute, Wuhan 430205, China
[2] School of Naval architecture and Ocean Engineering, Huazhong University of Science and Technology, Wuhan 430074, China

Abstract. The response of structure under shock is very important. For sandwich structures, the foam of other material with good energy absorbing capacity is an effective way to enhance the ability to resist external shock. In this work, numerical shock analysis is conducted for double-layered plate with foamed aluminum. The deformation process is firstly studied in case of initial shock velocity of 8.9m/s, then comparative analysis is conducted on the influence of different shock velocity on structural response.

1. Introduction

Structures in engineering would be strongly damaged, when it is shocked by explosion other flying pieces. Accurate prediction for deformation process of structure under chock load is very important and it will lay good foundation for shock resistance design of structures.
Recently, approaches applied to study the shock characteristics of structures are mainly numerical simulation method and experiment method.
Sandwich structures have wild applications in aerospace, ship structures, submarine, and so on. Many researchers have made some work on mechanism and analysis method for sandwich structures[1-3]. Besides, Omar et.al [4] studied deformation response of sandwich beam with foamed metal under three-point flexural load, the maximum stress is obtained and the failure mode is also discussed. Colombo et.al [5] analyzed the nonlinear characteristics of beton sandwich beam with three different core materials under four-point flexural load. Kobayashi et.al [6] investigated the response and failure mode for sandwich plate with circular core and honeycomb core under local repeating compressive load. But, research on the deformation process and failure for doubled-layered plate with foam core is few.
In this work, in order to get clear the response process of double-layered plate with foamed aluminum, numerical simulation is carried out with software ABAQUS. The deformation process is studied and the influence of shock speed on structural response is also conducted.

2. Numerical model

The Dynamic/Explicit module is applied. Solid model is constructed for structure and the drop hammer is rigid. Tie link is applied for connect between ring-stiffer and up/down plates.
Structural model. Structural model is shown in figure 1, and so is the mesh and boundary condition. The element is C3D8R, initial speed of drop hammer is calculated with different height. For example, initial speed of drop hammer is 6.5m/s, 8.9m/s, 11.5m/s in cases of height with 2.1m, 4m, 6.7m, respectively.

Fig. 1 Structural model and boundary condition for doubled-layered plate with foamed core

Material. Structure is 304 stainless steel, and the JOHNSON-COOK model is also applied. The mass of drop hammer is 362kg. Compressive foam model is applied for foamed core material, the density is 550 kg/m^3, Possion's ratio is 0.3, elastic modulus is 300MPa.
Test point. In order to output acceleration and displacement for comparative analysis, three different points are selected both for up and down plates, as is shown in figure 2.

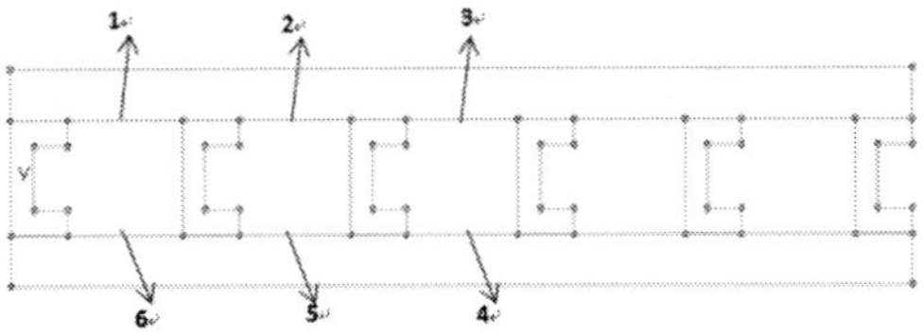

Fig. 2 Test points for numerical simulation

* Corresponding author: lizi@ustc.edu.cn

3. Numerical results analysis

3.1 shock process analysis

In the case of height 4m for drop hammer, the shock velocity is 8.9m/s, the shock process is plotted in Fig.3, and it's introduced in detail as below.

A) Drop hammer contact with the up plate, stress for up plate and ring-stiffer will be born, and the stress for up plate is relative bigger.

B) Drop hammer keeps dropping, stress for up plate and ring-stiffer will increase, and deformation is observed, thus the foamed core will be compressed.

C) Deformation will be observed for ring-stiffer, but it has no obvious damage, resistance ability for up plate will reach maximum, and its will be torn. At the same time, foamed core will separate from ring-stiffer.

D) With the increase of load, crack for the up plate will grow bigger and bigger, deformation of ring-stiffer will also be bigger and bigger. And the foamed core will more and more compressed.

E) Crack of up plate increases, and crack is also noted for joint between plate and ring-stiffer, failure is born in left ring-dtiffer, and the foamed core will crush.

F) Drop hammer keep dropping, crack of up plate and ring-stiffer will be bigger, while the foamed core will be densification and will be completely damaged. The total deformation will be increase.

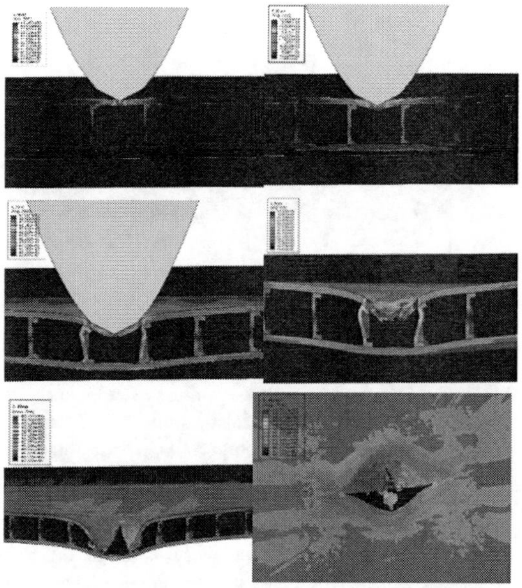

Fig. 3 Deformation process of double-layered plate with foamed aluminum under initial shock velocity of 8.9m/s

In the total process, max deformation for up plate is 24.5mm, the stress and strain is shown in figure 4-5. It can be concluded that concentration is very obvious , and the absorbed energy for structure is 4.9×103J.

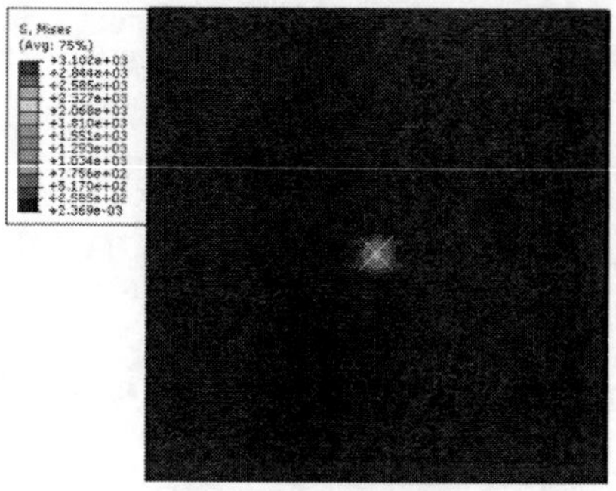

Fig. 4 Stress of double-layered plate with foamed core

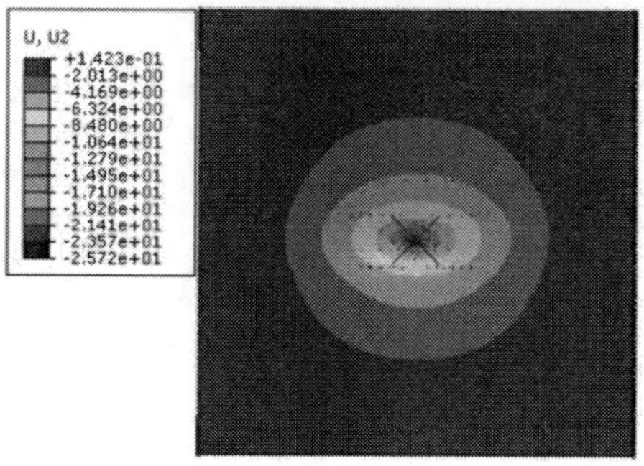

Fig. 5 Deformation of double-layered plate with foamed core

3.2 comparative analysis in case of different shock velocity

Comparative analysis is conducted for deformation and failure of double-layered plate with foamed core under three different shock velocities, the results are shown in figure 6.

In the case of relative low shock velocity, 6.5m/s, local plastic deformation is observed for structures. The shock energy is absorbed with light crack and local plastic deformation of up plate, and the foamed core and ring-stiffer are lightly compressed.

While the shock velocity increase to be 8.9m/s, crack of up plate is more obvous, and crack length and area increases; the foam core near the drop hammer is strongly densification and crushes; failure for the ring-stiffer is born. Shock energy absorption increases a little, due to the crack of up plate and densification of foamed core.

While the shock velocity is 11.5m/s, structural failure is more obvious, since the action of drop hammer is more significant. Hammer into the up plate can be noted, and the crack area increases. The densification of foamed core near drop hammer is much more serious, and it crushes and become failure under compressive load. With comparison with the case of 8.9m/s, carrying capacity of ring-stiffer in case of 11.5m/s increases a little, which will result in the obvious increase of plastic deformation of down plate.

https://doi.org/10.1051/matecconf/202338001007

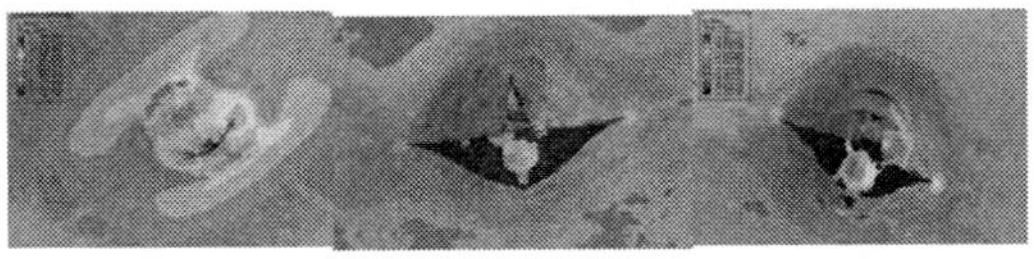

(a) 6.5m/s; (b) 8.9m/s; (c) 11.5m/s

Fig. 6 deformation of doubled-layered plates under different shock velocities

4. Conclusion

In this work, numerical shock analysis is conducted for double-layered plate with foamed aluminum. The deformation process is firstly studied in case of initial shock velocity of 8.9m/s, then comparative analysis is conducted on the influence of different shock velocity on structural response.

Conclusions obtained in this work will lay foundation for future work for making good shock test plan for double-layered plate with foamed aluminum.

Acknowledgements

We sincerely thank the financial support from the Hubei Provincial Natural Science Foundation (No. ZRMS2020002231) and the Young Top-notch Talent Cultivation Program of Hubei Province.

References

1. NOOR A K, BURTON W S, BERT C W. Computational models for sandwich plates and shells[J]. Applied Mechanical Review . 1996,49:155–199.
2. ABRATE S. Localized impact on sandwich structures with laminated facings[J]. Applied Mechanical Review. 1997,50: 69–82.
3. VINSON J R. Sandwich structures[J]. Applied Mechanical Review. 2001, 54:201–214.
4. Omar M Y, Xiang C, Gupta N, et al. Syntactic foam core metal matrix sandwich composite under bending conditions[J]. Materials & Design, 2015, 86:536-544.
5. Colombo I G, Colombo M, Prisco M D. Bending behaviour of Textile Reinforced Concrete sandwich beams[J]. Construction & Building Materials, 2015, 95:675–685.
6. Kobayashi Y, Kishimoto Y, Ohtsuka T, et al. Local compression fatigue property of roll core sandwich panel[J]. Mechanical Engineering Journal, 2015, 2(1):14-00322-14-00322.

Study on preparation and slow release properties of Magnesium-Al hydrotalcite with alanine intercalation

Simin Feng [1, 2, 3, *], Qing Liu [1, 2, 3], Jing Wei [1, 2, 3]

[1] Shaanxi Provincial Land Engineering Construction Group Co., Ltd., Xi'an, Shaanxi, China
[2] Institute of Land Engineering and Technology, Shaanxi Provincial Land Engineering Construction Group Co., Ltd., Xi'an, Shaanxi, China
[3] Land Engineering Quality Testing of Shaanxi Land Engineering Construction Group Co., Ltd., China

Abstract. The precursors of Mg-Al hydrotalcite (Mg/Al-LDHs-NO_3) were prepared by low saturation coprecipitation method (constant pH method). In addition, alsine-intercalated magnesium-aluminum hydrotalcite (LDHs-Ala) was synthesized by low saturation coprecipitation (constant pH) and nucleation/crystallization isolation methods when the pH value was higher than the isoelectric point of alanine. The release properties of alanine intercalated hydrotalc in simulated gastric fluid (hydrochloric acid solution with pH=1.5) and intestinal fluid (phosphate solution with pH=7.4) were measured by spectrophotometer in vitro. The results showed that LDHs-Ala had good slow-release properties in simulated intestinal fluid and gastric fluid in vitro. Then sodium alginate was used to coated alanine intercalated hydrotalc for slow release in simulated intestinal fluid and simulated gastric fluid. It can be concluded that the coated intercalated hydrotalc has a low slow release rate in gastric fluid, but has a good slow release performance in simulated intestinal fluid. It can be concluded that the coated intercalated hydrotalc is expected to be a intestinal targeted release agent, which can be applied in biomedicine.

Key words: hydrotalcite; Alanine; Slow release; Intercalation; coating.

1. Introduction

Hydrotalcite is a kind of anionic intercalated composite with layered structure. Anionic intercalation composites have a certain amount of positive charge on the laminates, so some anions can enter the laminates and act on the laminates with a certain weak chemical force, such as van der Waals force and hydrogen bond, so as to form anionic intercalation composites. There are natural hydrotalcite in nature, but also can be artificial synthesis in the laboratory, and the operation is simple, synthesis methods are more kinds, including coprecipitation method, roasting reduction method, ion exchange method, etc. The magnesium ions on the hydrotalcite laminates can be replaced by other metal ions in the same crystal. The interlayer anions are exchangable, so the hydrotalcite compounds with different functions can be synthesized, which has a good application prospect in the field of functional polymers.

In the past, organic matter was used as the skeleton material in the drug slow release system, but there may be shortcomings such as poor mechanical strength, poor thermal stability and secondary pollution of organic solvent. In view of the excellent properties of hydrotalcite, people gradually use hydrotalcite as a drug carrier to form supramolecular structure system. The drug intercalated hydrotalcite obtained by intercalation assembly can improve the thermal stability, acid-base resistance and drug slow release performance of hydrotalcite, and the intercalation assembled hydrotalcite compounds have good biocompatibility and will not produce rejection reaction in human body, so it has been widely concerned by researchers. This topic through the in-depth study of hydrotalcite, according to the special properties of hydrotalcite, amino acids and hydrotalcite intercalation assembly under certain circumstances, expect to prepare a new type of amino acid drug sustained-release agent, so as to study the slow-release performance of amino acid molecules in human body and achieve the targeted release of amino acids. This type of drug release system can be targeted to certain organs to achieve specific efficacy and avoid harming other organs.

2. Proposal of research questions

Amino acid drugs have some limitations, such as poor acid and alkaline resistance and poor absorption. In view of the advantages of drug intercalated hydrotalcite in the field of drug slow release, I used hydrotalcite as a carrier and inserted alanine into hydrotalcite to study the slow release performance of alanine intercalated hydrotalcite.

* Corresponding author: 871787254@qq.com

3. Experiment

3.1 Pharmaceutical reagents, instruments and equipment

(1) Pharmaceutical reagent: Magnesium nitrate ($Mg(NO_3)_2 \cdot 6H_2O$), Tiantian Chemical Reagent Factory, Dongli District, Tianjin; Aluminum nitrate ($Al(NO_3)_3 \cdot 9H_2O$), Haijing Fine Chemical Plant, Hebei District, Tianjin; Sodium hydroxide (NaOH), Tianjin Hongyan Chemical Reagent Factory; Sodium dihydrogen phosphate ($NaH_2PO_4 \cdot 2H_2O$), Tianjin Chemical Reagent Factory No. 3; Hydrochloric acid (HCl), Tianjin Chemical Reagent Factory No. 3; Alanine ($C_3H_7O_2N$), Shanghai reagent Factory No. 3; Ascorbic acid ($C_6H_8O_6$), Tianjin Chemical Reagent Factory No. 3; Ninhydrin ($C_9H_4O_3$), Shanghai Reagent Factory No. 3; Pyridine (C_5H_5N), third sea reagent plant.

(2) Equipment: Multi-frequency chemical reaction generator (SC-III type), Chengdu Jiuzhou Ultrasonic Technology Co., LTD. Vacuum Drying Oven (DZX-3 Type), Shanghai Fuma Experimental Equipment Co., LTD. Electronic Constant Temperature Stainless Steel Water Bath (HS-1S), Shanghai Guangdi Instrument Equipment Co., LTD. pH meter (PHS-3D), Nanchang Gaoteng Technology Co., LTD.; Electronic balance (Type AG135), Mettler, Switzerland.

3.2 Preparation of class hydrotalcite and generation of coated sample

3.2.1 Mg/Al-LDHs-NO₃ hydrotalcite was prepared by constant pH method

Weighing 20.51g(0.08mol) magnesium nitrate and 15.01g(0.04mol) aluminum nitrate in A certain amount of boiling deionized water is solution A; About 9.6g(0.24mol) of sodium hydroxide was weighed and dissolved in boiled deionized water, denoted as solution B. Solution B and solution A were dripped out at a certain rate, and then added to a large beek to gradually produce milky white precipitation. The pH value of the solution was always adjusted to be between 9 and 10. After dripping, the solution was reflow stirred for 4h at 70℃, the resulting slurry was crystallized for 5h, pumped out and washed with deionized water until neutral. Dry in oven at 65℃ for 24h. Grind and store in a drying oven.

3.2.2 Preparation of alanine intercalated Mg-Al hydrotalcite by constant pH method

Weighing 20.51g(0.08mol) magnesium nitrate and 15.01g(0.04mol) aluminum nitrate in A certain amount of boiling deionized water is solution A; Weighing alanine (Ala) 3.56g(0.04mol) dissolved in a certain amount of boiling deionized water is the solution. Under intense agitation and N2 protection, solution A was added to solution B at a constant rate, keeping the pH equal to 9~10 during the drip process. At pH value, a certain concentration of sodium hydroxide solution was used to adjust the drip, and then the slurry was reflow and stirred at 70℃ for 4h. The resulting slurry was liquid crystalline for 5h, then filtered and washed to neutral with deionized water. Dry in oven at 65℃ for 24h. Grind and store in a drying oven. Denoted as sample LDHs-Ala-[1].

3.2.3 Preparation of alanine intercalated Mg-Al hydrotalcite by nucleation crystallization isolation method

Weighing 20.51g(0.08mol) magnesium nitrate and 15.01g(0.04mol) aluminum nitrate in A certain amount of boiling deionized water is solution A; Weighing alanine (Ala) 3.56g(0.04mol) dissolved in a certain amount of boiling deionized water is the solution. Under intense agitation and N2 protection, solution A was added to solution B at a certain rate, and the pH was always maintained between 9 and 10 during the drip process. The pH was adjusted with sodium hydroxide solution. After drip adding, it was put into a multi-frequency chemical reactor for 40min. Finally, the solution was inverted into a three-necked flask for reflux stirring for 4h. Finally, it was crystallized and filtered, washed to pH less than 8, and dried in a 65℃ oven for 24h. Grind and store in a drying oven. It is denoted as LDHs-Ala-[2].

3.2.4 Generation of sodium alginate coated hydrotalcite

Solution A was obtained by dissolving 4g sodium alginate in a certain amount of deionized water and adding 0.4g or 0.27g alanine intercalated hydrotalcite LDHs-Ala-[2]. 4.4g $CaCl_2 \cdot 2H_2O$ was dissolved in 100mL deionized water to obtain solution B. After agitation, the solution A was slowly dropped into the solution B. After titration, the solution was filtered, washed with water, and dried at 65℃ in the oven for 24h. After drying, use a mortar to grind the powder as LDHs-Ala(musculo-1:10) or LDHs-Ala(musculo-1:15).

3.3 Determination of slow-release properties

The retarded release property of alanine can be determined by ninhydrin chromogenic method. This is the use of ninhydrin and alanine can produce purple or blue complex under a certain pH condition, using 722 spectrophotometer to determine the absorbance of alanine.

3.3.1 Preparation of ninhydrin color developing solution

Mix 100mL anhydrous ethanol with a small amount of water, add 0.5g ninhydrin, add 1.5mL pyridine, and a small amount of ascorbic acid, shake well to form a color solution.

3.3.2 Drawing of alanine standard curve

1g alanine was accurately weighed and diluted with deionized water into a constant volume 1000mL volumetric bottle to prepare 1g/L alanine buffer solution. A certain volume of alanine mother liquor was absorbed respectively. The pH of 6 phosphate buffer solution was diluted into 25mL colorimetric tube, and alanine standard solution of different concentrations was prepared as 0mg/L, 20mg/L, 40mg/L, 60mg/L, 80mg/L, 100mg/L, 120mg/L, and 1mL color developing agent was added. Cover the stopper of the colorimetric tube and heat in a constant temperature water bath for 20min. The absorbance was measured on a 722 spectrophotometer at 570nm wavelength, and the standard working curve of alanine was drawn.

3.3.3 Sustained release experiment of alanine intercalation Mg-Al hydrotalcite

(1) Determination of the total amount of alanine inserted into hydrotalcite
Take 1g of alanine intercalation magnesium aluminum hydrotalc and stir it with 0.3mol/L hydrochloric acid continuously to dissolve it, and dilute it with distilled water into a 100mL volumetric bottle, then accurately measure 2mL of the solution and add it into a 25mL colorimetric tube. Follow the same procedure as the standard curve for color development and measurement of absorbance.
(2) Measurement of slow release properties
First, 100mL of simulated gastric fluid and simulated intestinal fluid were prepared, and then 0.5g each of alanine intercalated magnesia al hydrotalc LDHs-Ala-[1], LDHs-Ala-[2] and LDHs-Ala(coating) prepared by constant pH method and nucleation/crystallization isolation method were added, and stirred with magnetic stirring instrument. The temperature of the slow-release system was kept at (37±1)℃. Every once in a while, accurately measure 2mL of the solution, add it to the colorimetric tube, carry out color rendering in the same steps as the standard curve, and measure the absorbance. At the same time, 2mL hydrochloric acid solution with pH=1.5 or phosphate buffer solution with pH=7.4 were added into the simulated sustained release system in vitro. Finally, the sustained release rate at different time can be obtained by calculation.
Reference experiment: Weighing 0.110g Mg/Al-LDHs-NO_3 and 0.100g Ala(NO_3^-:Ala=1:1), the two samples were ground evenly in a mortar. The cumulative slow release of alanine (Ala) in hydrochloric acid solution with pH=1.5 and phosphate solution with pH=7.4 was determined by the above method.

4. Results and discussion

4.1 Standard curve of alanine

According to the experimental part, the standard curve of alanine (Ala) was drawn, and the equation of the curve was as follows: Y=-0.00393+0.00602X, R2=0.99966.

4.2 Determination of the release rate in simulated human gastric juices

According to the experimental part, the propyl acid intercalation magnesium-aluminum hydrotalcite LDHs-Ala-[1], LDHs-Ala-[2] and the physical mixture of magnesium-aluminum nitrate hydrotalcite precursors and alanine prepared by constant pH method and nucleation isolation method were released in hydrochloric acid solution with pH=1.5 for 3h, and the release curve was obtained.
As can be seen from Figure 4-1, in the simulated human gastric fluid with pH=1.5, the sample release rate is particularly fast in the first 5 to 10min, which may be because hydrotalcite is layered dihydroxyl hydroxide, which is unstable under acidic conditions. The lamellar may be damaged and partially dissolved and released alanine, and then the release rate gradually slows down as time goes by. In the sustained release experiment of LDHs-Ala-[2], there are 5, 10, 30, 80, 180 In min, the release rates were 50.45%, 61.38%, 72.03%, 85.02% and 92.88%, respectively. The slow release process was a slow release, which may be caused by the exchange of chloride ions in solution with alanine between LDHs-Ala-[2] layers, and chloride ions entered the LDHs layers. However, when alanine enters the solution, the ion exchange rate is fast at the beginning stage, and then decreases gradually after reaching the maximum. Hydrotalc was destroyed in the middle plate of hydrochloric acid solution and the slow release rate of alanine intercalation of hydrotalc was increased. By comparing the two methods for the synthesis of alanine intercalation hydrotalcite, it can be concluded that the release amount of LDHs-Ala-[2] is more than that of LDHs-Ala-[1] at 180 min, indicating that the nucleation crystallization isolation method is easier to insert alanine into the hydrotalcite laminate.

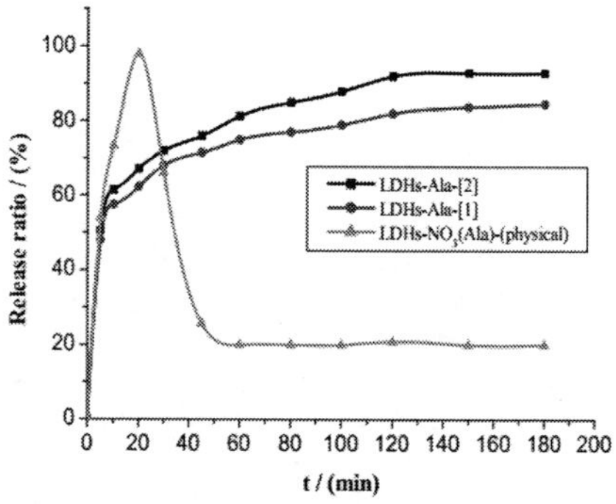

Figure 1. alanine slow-release curve in simulated human gastric fluid

It can be seen from Figure 1 that the release process of physical mixture of magnesium aluminum nitrate hydrotalcite precursor and alanine in simulated human gastric fluid shows that the release rate of amino acid increases and decreases rapidly in 5-20min, which may be due to the poor acid resistance of alanine and its partial decomposition in hydrochloric acid solution. It can be concluded that the stability of alanine can be improved when alanine is inserted into hydrotalcite, and the release performance is relatively good.

4.2.1 Determination of release rate in simulated human intestinal fluid

LDHs-Ala-[1], LDHs-Ala-[2] and the physical mixture of precursors and alanine of propyl acid intercalated magnesium and aluminum hydrotalcite prepared by constant pH method and nucleation isolation method, as well as magnesium and aluminum nitrate hydrotalcite, were released in phosphate solution at pH=7.4 for 3 hours, and the release curve was obtained.

As can be seen from Figure 2, in the simulated human intestinal fluid with pH=7.4, the sample is basically released completely in about 30min, and the release rate finally persists at about 98.8%. The release of alanine intercalation hydrotalc synthesized by two different methods in the simulated human intestinal fluid does not occur suddenly from beginning to end, but it is a slow and continuous process. In the sustained release experiment of LDHs-Ala-[2], the release rates at 5, 10, 30, 80 and 180min were 34.94%, 48.55%, 60.40%, 77.03% and 84.99%, respectively. The release process in intestinal fluid is actually the result of the ion exchange between LDHs-Ala and phosphate in intestinal fluid, that is, phosphate ions in the sustained-release solution enter the interlaminates of hydrotalc, and alanine anions between the interlaminates of hydrotalc enter the sustained-release solution. This process is a slow and continuous process. Compared with physical mixtures, alanine intercalated hydrotalcite has better slow release properties.

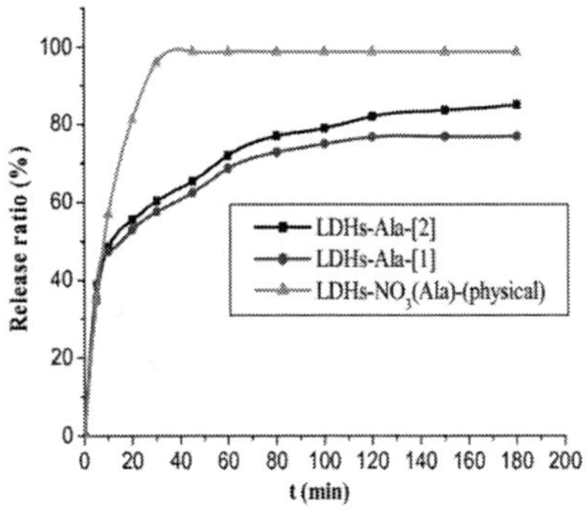

Figure 2. alanine slow-release curve in simulated human intestinal fluid

Similar to the simulated human gastric fluid, the slow release performance of the simulated human intestinal fluid is also that the alanine release of the alanine intercalated hydrotalcite prepared by nucleation crystallization isolation method is more and longer, which once again proves that the intercalated hydrotalcite prepared by nucleation crystallization isolation method has better performance.

4.2.2 Determination of the release rate of coated samples in simulated human gastric fluid intestinal fluid

The propyl acid intercalated magnesium aluminum hydrotalc LDHs-Ala-[1] and LDHs-Ala-[2] prepared by constant pH method and nucleation isolation method were coated with sodium alginate at a mass ratio of 1:10 and 1:15, respectively, and then released into the simulated intestinal fluid with pH=7.4 and the simulated gastric fluid with pH=1.5 for 2h to obtain the release curve.

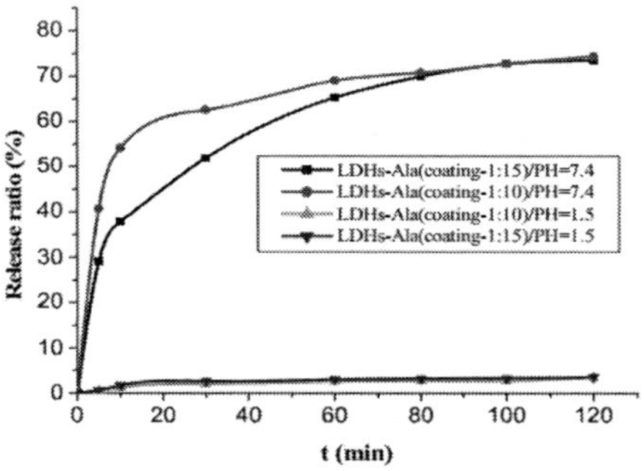

Figure 3. Slow release curve of coated samples in simulated human gastric and intestinal fluids

The samples coated with sodium alginate, LDHs-Ala (musculo-1:10) or LDHs-Ala (musculo-1:15), were put into simulated human gastric fluid and intestinal fluid respectively for the determination of slow-release properties, and released for two hours. As can be seen from Figure 3, alanine release in simulated human gastric juices is rarely less than 3%. In the simulated human intestinal fluid, alanine can release LDHs-Ala (musculo-1:15) slowly, and the release rates at 5, 10, 30, 80 and 180 min are 29.12%, 37.89%, 51.96%, 70.00% and 73.49%, respectively. The release rates of LDHs-Ala (coating-1:10) at 5, 10, 30, 80 and 180min were 40.74%, 54.21%, 62.63%, 72.8% and 74.38%, respectively. It can be concluded that the release performance of coated amino acid intercalated hydrotalc in human simulated intestinal fluid is better, and it can achieve sustained slow release in a long time, and the acid resistance is significantly improved. Therefore, the coated alanine intercalated hydrotalc is expected to be a targeted release agent of intestinal fluid, which can be widely used in biomedicine.

5. Conclusion

In this paper, the precursors of nitrate hydrotalcite were prepared by low saturation coprecipitation method (constant pH method), and alanine intercalated magnesium aluminum hydrotalcite was prepared by low saturation coprecipitation method (constant pH method) and nucleation/crystallization isolation method respectively. Then, the slow-release properties of alanine intercalated MAl-Al hydrotalc were measured in vitro by simulating human intestinal fluid (phosphate solution with pH=7.4) and human gastric fluid (hydrochloric acid solution with pH=1.5). The results were as follows: Compared with the physical mixture of hydrotalcite and alanine, the hydrotalcite assembled by alanine intercalation can effectively improve the acid resistance of alanine, and has good slow-release performance in the simulated human intestinal and gastric fluid systems. LDHs-Ala coated with sodium alginate is basically not released in simulated human gastric fluid, but has good slow-release performance in simulated human intestinal fluid, that is, it can be released slowly and persistently for a long period of time. Therefore, amino acid intercalation hydrotalc is expected to become a targeted intestinal release agent and be applied in the field of biomedicine.

Acknowledgments

This work is supported, in part, by Innovation Capability Support Program of Shaanxi (Program No.2021PT-053).

References

1. Trikeriotis M, Ghanotakis D F. Intercalation of hydrophilic antibiotics in layered double hydroxides [J]. International Journal of Pharmaceutics, 2007, 332(1-2): 176~184.
2. Shaoyang Z, Quanyao L, Guoli F, et al. Characteristics and mechanism of sorption of organic contaminant onto sodium dodecyl sulfate modified Ca-Al layered double hydroxides [J]. Chinese Science Bulletin, 2010(12): 3431~3436.
3. Cllaudla R G, Qin C A S, Ana M F, et al. Immobilization of ibuprofen and copper-ibuprofen drugs on layered double hydroxide [J]. J Phys sci, 2005, 94(5): 1135~1147.
4. Hussein M Z,Yahaya A H, Zainal Z, et al. Nanocomposite-based controlled release of an herbicide: 2,4-dichlorophenoxyacetateincapsulatedin zincalumin-ium intercalation layered double hydroxide[J]. Sci Technol Adv Mater, 2005, 25(6): 956~962.
5. Valeria A, Giuseppe F, Giuliano G. Intercalation Compounds of Hydrotalcite-like Anionic Clays With Anti-inflammatory Agents II: Uptake of Diclofenac for a Controlled Release Formulation [J]. AAPS Pharmscitech, 2002, 23(13): 1~6.

Design of a NC machine tool for machining that is easy to change cutting tools

Zhiqiang Wan, Qingqing Zhang

Anhui Technical College Of Mechanical and Electrical Engineering, Wuhu, Anhui, China

Abstract. A CNC machine tool for machining designed to facilitate the replacement of cutting tools includes a workbench, a motor and a hydraulic cylinder. The inner bearing of the workbench is connected with a rotating rod, the outer key of the rotating rod is connected with a half gear, and the output end of the motor is connected with the lower end of the rotating rod. The CNC machine tool for machining, which is convenient for replacing cutting tools, realizes the linkage between the cutting tool body and the adjusting gear by using the tooth block, and drives the adjusting rod to move by using the rotation of the screw rod to adjust the position between the pawl and the ratchet teeth. The rotation of the adjusting gear is limited by using the pawl and the ratchet teeth, so as to lock the cutting tool body and quickly disassemble and install the cutting tool body.

Keywords: CNC machine , cutting tools, machining, replacement.

1. Introduction

Machining is the process of changing the shape, size, position and nature of the production object to make it become a finished product or semi-finished product. According to the differences in processing methods, it can be divided into cutting and pressure processing. Cutting CNC machines are often used in the cutting process of machining. Cutting CNC machines can be used to cut the workpiece quickly and conveniently.

However, the existing CNC machine tools for machining still have shortcomings in the process of use, such as not being able to conveniently disassemble and replace the cutting tools, being less convenient to use, not being convenient to maintain the cutting tools, and not being able to quickly update the cutting tools in case of cutting tool failure in the processing process, which affects the subsequent processing efficiency.

The purpose of this design is to provide a CNC machine tool for machining that is easy to replace cutting tools, so as to solve the above-mentioned problem that the existing CNC machine tools for machining in the market can not easily disassemble and replace the cutting tools, and can not quickly update the cutting tools when the cutting tools fail during processing.

2. Technical Design Scheme

A CNC machine tool for machining that is easy to replace cutting tools includes a workbench, a motor and a hydraulic cylinder. The inner bearing of the workbench is connected with a rotating rod, and the outer key of the rotating rod is connected with a half gear. The output end of the motor is connected with the lower end of the rotating rod. The right side of the half gear is meshed with a full gear, and the full gear key is connected with the outer side of the lower end of the tool holder. The tool holder bearing is connected above the workbench, And a mounting bracket is installed on the outer side above the tool bracket, and a cutting tool body is arranged inside the mounting bracket. The CNC machine tool for machining, which is convenient for replacing cutting tools, realizes the linkage between the cutting tool body and the adjusting gear by using the tooth block, drives the adjusting rod to move by using the screw rod rotation, adjusts the position between the pawl and the ratchet tooth, and limits the rotation of the adjusting gear by using the pawl and the ratchet tooth, so as to lock the cutting tool body and quickly disassemble and install the cutting tool body.

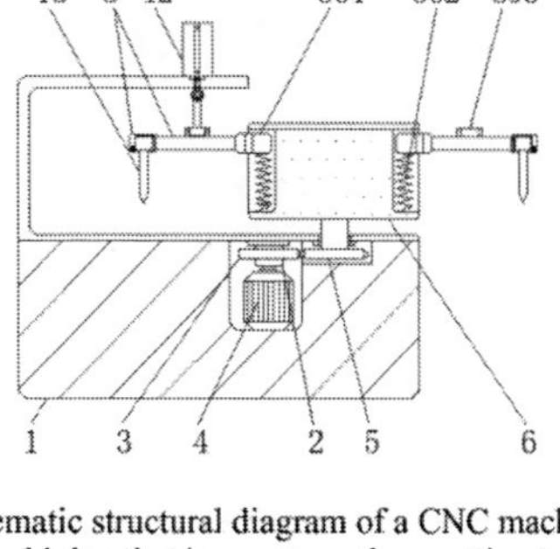

Fig. 1 Schematic structural diagram of a CNC machine tool for machining that is easy to replace cutting tools

1. Workbench; 2. Rotating rod; 3. Half gear; 4. Motor; 5. Full gear; 6. Knife rest; 7. Adjusting lever; 8. Mounting bracket; 801. Slider; 802. Compression spring; 803. Card holder; 9. Screw rod; 10. Tooth block; 11. Chuck; 12. Hydraulic cylinder; 13. Cutting tool body; 14. Adjusting gear; 15. Ratchet teeth; 16. Pawl; 17. Rotating shaft; 18. Scroll spring.

3. Example given to illustrate

A CNC machine tool for machining that is convenient for replacing cutting tools, as shown in Fig. 1, comprises a table 1, a rotating rod 2, a half gear 3, a motor 4, a full gear 5, a tool rest 6, an adjusting rod 7, a mounting bracket 8, a slider 801, a compression spring 802, a chuck 803, a screw rod 9, a tooth block 10, a chuck 11, a hydraulic cylinder 12, a cutting tool body 13, an adjusting gear 14, ratchet teeth 15, pawls 16, a rotating shaft 17 and a scroll spring 18. The internal bearing of the table 1 is connected with the rotating rod 2, The outer side of the rotary lever 2 is keyed to a half gear 3, the output end of the motor 4 is connected to the lower end of the rotary lever 2, the right side of the half gear 3 is meshed to a full gear 5, and the full gear 5 is keyed to the outer side of the lower end of the tool holder 6. The tool holder 6 is connected to the upper side of the table 1 by a bearing, and a mounting bracket 8 is installed on the outer side above the tool holder 6. A cutting tool body 13 is installed inside the mounting bracket 8, and a tooth block 10 is installed on the outer side of the cutting tool body 13, The outer side of the gear block 10 is meshed and connected with an adjusting gear 14, and the inner side of the adjusting gear 14 is connected with a ratchet tooth 15. The outer side of the ratchet tooth 15 is fitted with a pawl 16, and the pawl 16 is welded and connected to the outer side of the rotating shaft 17. The rotating shaft 17 is connected with the adjusting rod 7 through a scroll spring 18. The screw 9 is connected to the internal thread of the adjusting rod 7, and the screw 9 is connected to the mounting frame 8 through bearings. The hydraulic cylinder 12 is bolted to the upper part of the workbench 1, And a chuck 11 is attached to the lower end of the hydraulic cylinder 12.

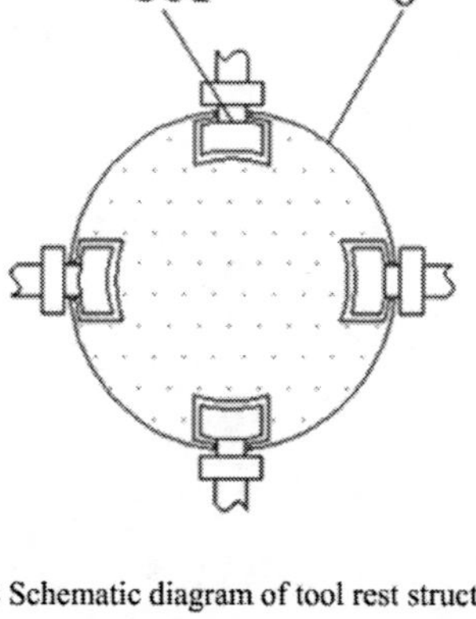

Fig. 2 Schematic diagram of tool rest structure

The half gear 3 and the full gear 5 have the same size, and the number of teeth of the half gear 3 is one fourth of the number of teeth of the full gear 5. This design can drive the tool holder 6 to rotate a quarter of the surface at a time, so that the position of the mounting bracket 8 can be easily and quickly replaced.

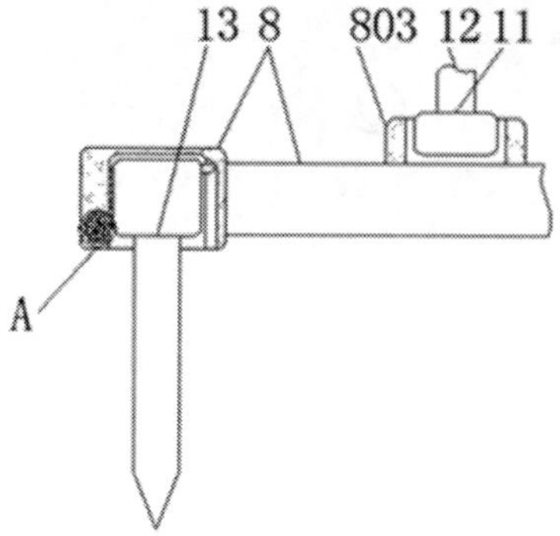

Fig. 3 Structural diagram of the connection between the mounting bracket and the cutting tool body

The tool rest 6 and the table 1 constitute a rotating mechanism, and the top section of the tool rest 6 is a circular structure, and the mounting brackets 8 are distributed at equal angles on the outer side of the tool rest 6. This design can stably control the rotation of the tool rest 6 and adjust the positions of the mounting brackets 8 in different directions;

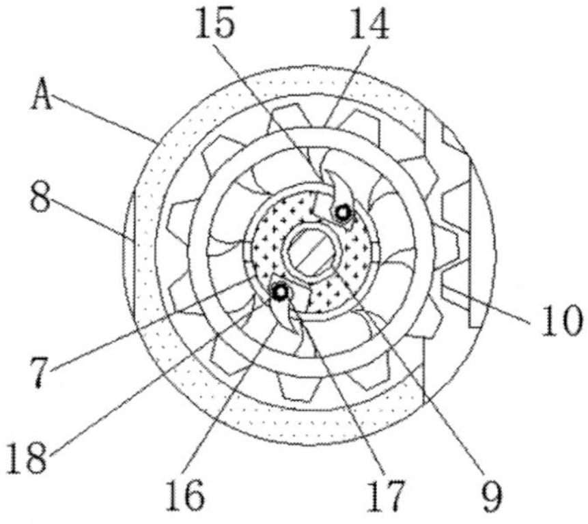

Fig. 4 Enlarged structural diagram at a in Fig. 3

The mounting frame 8 includes a slider 801, a compression spring 802 and a holder 803. The outer end of the mounting frame 8 is fixedly mounted with a slider 801, and the lower side of the slider 801 is connected with a compression spring 802. At the same time, the upper side of the mounting frame 8 is welded with a holder 803. This design can control the up and down reciprocating movement of the mounting frame 8, thereby controlling the up and down reciprocating movement of the cutting tool body 13, and cutting the workpiece with the cutting tool body 13.

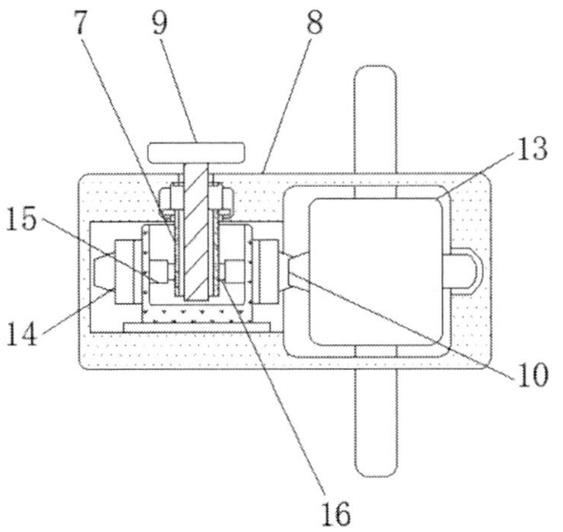

Fig. 5 Structural diagram of mounting bracket

The top section of the slider 801 has an I-shaped structure, and the slider 801 forms an elastic structure with the tool rest 6 through the compression spring 802. This design can ensure the stable vertical movement of the mounting bracket 8 and prevent the mounting bracket 8 from being disconnected from the tool rest 6;

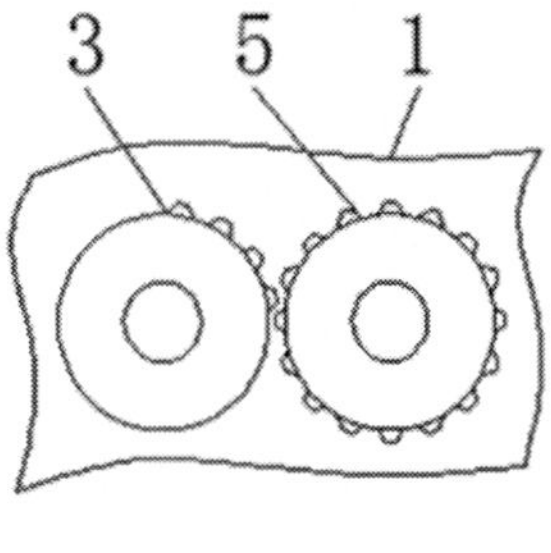

Fig. 6 Structural diagram of half gear and full gear connection

The tooth blocks 10 are equally spaced at the joint of the cutting tool body 13 and the adjusting gear 14, and the center line of the adjusting gear 14 and the center line of the adjusting lever 7 coincide with each other. The top section of the adjusting lever 7 is a "t" shape. At the same time, the adjusting lever 7 and the mounting bracket 8 are slidably connected. This design can realize the linkage between the cutting tool body 13 and the adjusting gear 14. By locking the adjusting gear 14, the cutting tool body 13 can be locked.

4. Conclusion

Compared with the prior art, a CNC machine tool for machining that is easy to replace cutting tools has the following advantages:
(1) The gear block is used to realize the linkage between the cutting tool body and the adjusting gear, and the screw rod is used to drive the adjusting rod to move, adjust the position between the pawl and the ratchet tooth, and limit the rotation of the adjusting gear by the pawl and the ratchet tooth, so as to lock the cutting tool body and quickly disassemble and install the cutting tool body;
(2) The half gear rotates to drive the full gear to rotate by a quarter, so as to control the tool holder to rotate by a quarter, adjust the position of the mounting bracket in different directions, and update the cutting tool body. This design will not affect the operation of the CNC machine tool, and can easily replace the cutting tool body.

Acknowledgements

Supported by the scientific research project of Anhui Provincial Department of Education (KJ2021A1512, KJ2021A1515).

References

1. Li Yulan. Geometric accuracy measurement of CNC machine tools [M]. Beijing: China Machine Press, 2014.
2. Yang Jianguo. Integrated error compensation technology and application of CNC machine tools [M]. Shanghai: Shanghai Jiaotong University, 1998.
3. Zheng dianmo, Lin Hu, bu Xiaofei, Liu minglie. Design of general kinematic model of five axis machine tool [J]. Small microcomputer Systems, 2010, 31 (10): 1965-1969.
4. Fan Shutian, Yang Weiping. Research on RTCP function of double turntable five axis machine tool [J]. Manufacturing technology and machine tool, 2009 (12):74-77.
5. Jiang Zhong, Ding Jiexiong, Wang Wei, et al. Dynamic error tracing method of five axis NC machine tool based on RTCP function [J]. Machinery Journal of engineering, 2016, 52 (7): 187-195.

MATEC Web of Conferences **380**, 01010 (2023) https://doi.org/10.1051/matecconf/202338001010
MSMS 2023

Recycling of detoxification wastewater from Semen Armeniacae Amarum

Ke Liu, Yiyang Ren and Yongsheng Li *

School of Materials Science and Engineering, Shanghai Institute of Technology, Shanghai 201418, PR China

Abstract. Detoxification of Semen Armeniacae Amarum requires large amounts of water, and the detoxifying wastewater contains highly toxic hydrocyanic acid. This paper aims to explore the recovery of water after the removal of cyanide from it by chemical precipitation. The results show that when the dosage of ferrous sulfate is 1.76 times of the theoretical value, the cyanide removal in water can be reduced to about 2.70×10- 5mol/L, the hydrocyanic acid in Semen Armeniacae Amarum can be entirely removed when the treated water is recycled.

1. Introduction

Almonds and almond products are appearing more regularly in people's lives as their standard of living increases. Almonds are used in a growing number of table dishes, casual meals and plant-based beverages. Products made from apricots, such as apricot juice, apricot cake, apricot candy, are gaining popularity with consumers. However, amygdalin contained in Semen Armeniacae Amarum can be hydrolyzed to benzaldehyde, glucose, and hydrocyanic acid in the presence of amygdalinase or under acidic conditions 1. Hydrocyanic acid is highly toxic and can cause poisoning and even death in humans and animals. Thus, the detoxification of Semen Armeniacae Amarum is a necessary process. Currently, the processing and detoxification of almonds is mostly done by boiling water cooking, soaking, biological fermentation, etc. These processes are performed by hydrolyzing amygdalin under certain conditions to release cyanide. As a result, large amounts of cyanide will be produced in the detoxified wastewater. There is a high probability of contamination incidents if detoxifying wastewater is directly discharged or improperly treated2. Currently, treatment methods for cyanide-containing wastewater include acidification, recovery, chlorine oxidation, chemical precipitation, electrolysis and microbial treatment [3, 4], however, there has been little domestic and international research on the recovery of cyanide-containing wastewater from Semen Armeniacae Amarum detoxification. In the chemical precipitation method, ferrous sulfate is used as a precipitator, and the excess ferrous sulfate forms a bright blue Prussian blue precipitate in water with cyanide. In this process, cyanide will initially react with ferrous ions to form ferricyanide ions, and excess ferrous sulfate will eventually reduce the generated ferricyanide ions to bright blue ferricyanide precipitate [5, 6], This method is simple in operation, low in treatment cost, non-toxic, and widely used in the treatment of cyanide containing wastewater [7, 8, 9].
Since the wastewater from the detoxification of Semen Armeniacae Amarum does not contain additional heavy metal ions, it is possible to recover the wastewater from the detoxification of Semen Armeniacae Amarum. Based on the above background, the aim of this study is to determine the feasibility and conditions for the recovery of detoxified wastewater from Semen Armeniacae Amarum after removal of cyanide by ferrous sulfate.

2. Experiment

In this study, the cyanide content of a solution of Semen Armeniacae Amarum immersed in water was determined from the mass ratio of Semen Armeniacae Amarum to the deionized water. The amount of cyanide in the soaking solution was determined using the pyramidal method with isotretinoic acid, and the soaking solution was recovered by treating it with ferrous sulfate.

2.1 Reagents and Instruments

Ferrous sulfate heptahydrate, bitter almond, isonicotinic acid, pyrazolone, chloramine T, sodium hydroxide, acetic acid, potassium dihydrogen phosphate, disodium hydrogen phosphate, anhydrous ethanol, phenolphthalein, zinc acetate, tartaric acid, standard substance for analysis of cyanide in water (500mg/L), UV spectrophotometer, electronic analysis balance, electric mixer, oil bath, three necked flask, pipette, microinjector.

2.2 Method

2.2.1 Drawing of the standard curve of cyanide.

Prepare 1mg/L cyanide standard intermediate solution for cyanide composition analysis reference material

* Corresponding author: keliu3199198500@163.com

(500mg/L) in water, measure 0.00ml, 0.01ml, 0.05ml, 0.10ml, 0.50ml, 1.00ml, 2ml, 3ml, 4ml and 5ml cyanide standard intermediate solution respectively with pipette, and put them into 25ml volumetric flask, use isonicotinic acid precision method to determine cyanide content in sample tube, and draw standard curve diagram.

2.2.2 Poison removal by soaking bitter almonds.

Weigh an appropriate amount of bitter almond, add deionized water, prepare a bitter almond soaking solution with a mass ratio of 1:15, change the water every 12 hours at a certain temperature, store the soaking solutions, respectively, and determine the cyanide ion content in the soaking solution by using the isonicotinic acid precision method.

2.3 Recycling of the soaking solution.

2.3.1 The removal of cyanide in a soaking solution.

120 ml of the above soaking solution of bitter almonds was boiled for 12h, various doses of ferrous sulfate were added according to stoichiometric calculations, stirred under different conditions, filtered, and the cyanide ion content of the soaking solution was determined by the method of pyrazolone of isotretinoic acid.

2.3.2 Recycling soaking water.

Recycling of the soaking solution is explored by taking the treated soaking solution and reusing it under certain conditions.

3. Results and Analysis

3.1 Standard curve

The standard curves for cyanide are shown in figure. 1 below. Taking the cyanide concentration (mg/L) in the standard sample as the abscissa and the absorbance (Abs) as the ordinate, the standard curve y=0.1439x+0.0134 is obtained, where the correlation coefficient R^2=0.9992, and the absolute value of the two independent determination results obtained under repeated conditions does not exceed 10% of the arithmetic mean. It shows that the linear equation obtained has excellent linear correlation, and the test method is reliable and effective.

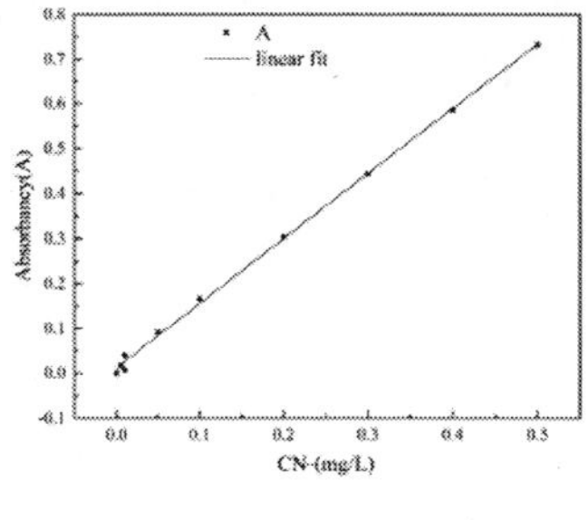

Figure 1. Cyanide (mg/L) standard curve

3.2 Poison removal by soaking almonds

3.2.1 Hydrocyanic acid concentration in different soaking solutions.

At room temperature, the temperature of ionized water is 20 °C, and the content of hydrocyanic acid in the replaced immersion solution is determined. The results are shown in Table 2: With the replacement of immersion solution, the content of cyanide in the water is different, and with the increase of replacement times, the concentration of cyanide in the immersion solution changes from 1.09×10^{-3}mol/L gradually approaches zero. The results showed that Semen Armeniacae Amarum was effective at releasing hydrocyanic acid through soaking, but produced large amounts of cyanide-containing wastewater while using large amounts of water.

Table 1. Cyanide concentration in different soaking solutions

Water change times	Cyanide Level (mol/L)
1	1.09×10^{-3}
2	4.35×10^{-4}
3	2.20×10^{-4}
4	6.31×10^{-5}
5	2.42×10^{-5}
6	0.538×10^{-5}
7	0.269×10^{-5}

3.2.2 The effect of temperature.

Amygdalin releases hydrocyanic acid under conditions of enzyme catalysis or weak acidity. The activity of the enzyme and the stability of the hydrocyanic acid are easily influenced by temperature. Therefore, it is necessary to soak for 12h at 20 °C, 25 °C, 30 °C, 35 °C, 40 °C, 45 °C respectively. The results are shown in Table 1. As the temperature increases, the concentration of cyanide in the immersed solution varies. When the temperature exceeds 35 °C, the concentration of hydrocyanic acid in the soaking solution starts to decrease. It shows that the stability of hydrocyanic acid is extremely sensitive to temperature under acidic conditions. When the temperature exceeds 35 wen °C, the hydrocyanic acid begins to volatilize. However, a proper increase in temperature can increase enzyme activity and improve the soaking effect. Therefore, we will soak Semen Armeniacae Amarum at 30 °C for detoxification.

Table 2. Detoxification Effect of Soaking at Different Temperatures

Soaking temperature (°C)	Cyanide Level (mol/L)
20	1.09×10^{-3}
25	1.12×10^{-3}
30	1.15×10^{-3}
35	0.932×10^{-3}
40	0.627×10^{-3}
45	0.495×10^{-3}

3.3 Recycling of soaking solution

It is known that ferrous ions can react with cyanide ions, and the generated complexes can be removed in the form of precipitation, which provides a theoretical basis for the removal of cyanide in water and the recycling of soaking solution.

3.3.1 Dosage of ferrous sulfate.

As mentioned above, for the cyanide ion of about 1.15×10^{-3}mol/L, theoretically about 158mg/L of ferrous sulfate is needed, but theoretically, ferrous ion is not enough to complexe with all cyanide ions [9], Therefore, at 20 °C, we use 158mg/L of ferrous sulfate as the starting dose of precipitator, and the interval between other doses is 30mg/L, add excessive ferrous sulfate, and use sodium hydroxide and dilute sulfuric acid to adjust the acidity and alkalinity of the solution before and after reaction, React under magnetic stirring for 30min, and filter after standing for 30min. The results are shown in Table 2: the theoretical addition of ferrous sulfate is not sufficient to entirely remove the cyanide. Within a certain range, the removal efficiency of cyanide increases continuously with the amount of ferrous sulfate. Treatment is best when ferrous sulfate is used in amounts up to 278 mg/L. Beckenn[10] suggested that in order to effectively remove cyanide, the molar ratio of Fe: CN should be 16, which is unnecessary, because this will lead to elevated consumption of the precipitant used. It is essential to note that an excess of ferrous sulfate also precipitates ferrous hydroxide, which plays a role in the precipitation process as an adsorbent.

Table 3. Results of ferrous sulfate precipitation at different doses

Initial Cyanide Level (mol/L)	Ferrous Sulfate Dose (mg/L)	Final Cyanide Level (mol/L)
1.15×10^{-3}	158	2.22×10^{-4}
1.15×10^{-3}	188	8.88×10^{-5}
1.15×10^{-3}	218	4.19×10^{-5}
1.15×10^{-3}	248	3.50×10^{-5}
1.15×10^{-3}	278	2.54×10^{-5}
1.15×10^{-3}	308	2.65×10^{-5}

3.3.2 Reaction time.

The reaction of ferrous sulfate with cyanide to produce precipitate is a chemical reaction process. Therefore, at 20 °C, the immersion solution with 278mg/L ferrous sulfate was stirred for 15min, 30min, 45min, 60min, 75min, and left for 30min. The results are shown in Table 4. For a reaction time of 30 min, the concentration of cyanide in the water reaches a minimum. When the stirring time exceeds 60 min, the cyanide concentration increases and it is possible that the mixing time is too long to destroy the resulting Prussian blue. Therefore, we set the reaction time as 30min.

Table 4. Cyanide concentration in soaking solution under different reaction time

Initial Cyanide Level (mol/L)	Ferrous Sulfate Dose (mg/L)	Mixing time (min)	Final Cyanide Level (mol/L)
1.15×10^{-3}	278	15	3.50×10^{-5}
1.15×10^{-3}	278	30	2.54×10^{-5}
1.15×10^{-3}	278	45	2.54×10^{-5}
1.15×10^{-3}	278	60	2.57×10^{-5}
1.15×10^{-3}	278	75	3.74×10^{-5}

3.3.3 The effect of temperature.

To explore the effect of temperature on the removal of cyanide by ferrous sulfate, an immersed solution of 278mg/L ferrous sulfate was stirred for 30min at various temperatures. The results are shown in Table 5: within a certain temperature range, the resulting Prussian blue precipitate is extremely stable and can effectively reduce the cyanide concentration in the immersed solution. Therefore, we react at 20 °C.

Table 5. Cyanide concentration in soaking solution at different reaction temperatures

Initial Cyanide Level (mol/L)	Reaction temperature (°C)	Final Cyanide Level (mol/L)
1.15×10^{-3}	20	2.54×10^{-5}
1.15×10^{-3}	25	2.58×10^{-5}
1.15×10^{-3}	30	2.54×10^{-5}
1.15×10^{-3}	35	2.46×10^{-5}

To sum up, at 20 °C, adding 278mg/L of ferrous sulfate, stirring reaction for 30min, the content of cyanide ion in the soaking solution is reduced to about 2.54×10^{-5}mol/L, the cyanide ion in the soaking solution is removed effectively, and the generated precipitation is stable, which makes it possible to recycle the soaking solution.

3.3.4 Reuse the soaking water.

Next, explore the recycling of the soaking solution: take 15g of unsinked Semen Armeniacae Amarum, add deionized water to form a mass ratio of 1:30, soak at 30 °C, switch the water every 12 hours, treat the first soaking solution under the best conditions, and then recycle it. The acidity and alkalinity of the solution were adjusted before and after reaction with sodium hydroxide and diluted sulfuric acid. It is essential to note that the amount of immersion solution lost due to each cyanide content determination, precipitation and filtration operation is supplemented by the same volume of deionized water. The results are given in Table 6 below: after applying ferrous sulfate to the immersed solution, the multiplexing is excellent and the concentration of cyanide in the immersed solution is continuously reduced. To investigate whether the hydrocyanic acid in the Semen Armeniacae Amarum had been completely removed, the

Semen Armeniacae Amarum that had been soaked seven times in a circular manner was again immersed in deionized water.

Table 6. Reuse of soaking solution

Number of cycles	Initial Cyanide Level (mol/L)	Ferrous Sulfate Dose (mg/L)	Final Cyanide Level (mol/L)
1	1.16×10^{-3}	278	2.57×10^{-5}
2	4.62×10^{-4}	112	2.57×10^{-5}
3	2.65×10^{-4}	64	2.54×10^{-5}
4	9.19×10^{-5}	23	2.57×10^{-5}
5	5.11×10^{-5}	13	2.65×10^{-5}
6	3.11×10^{-5}	10	2.67×10^{-5}
7	3.04×10^{-5}	10	2.81×10^{-5}

After the Semen Armeniacae Amarums have been soaked seven times in a cycle, they are re-soaked and cleaned with deionized water. The results are presented in Table 7 below. Semen Armeniacae Amarum is nearly free from hydrocyanic acid after one cycle of soaking. The traces of cyanide in the first soaking solution should be the same as those in the soaking solution of Semen Armeniacae Amarum adhered to, and may be effectually removed by soaking and cleaning. It can be seen that the soaking solution can be recovered after treatment with ferrous sulfate.

Table 7. Cyanide Concentration in Soaking Solution

Water change times	Cyanide Level (mol/L)
1	0.500×10^{-5}
2	0
3	0

4. Conclusion

After the cyanide in the detoxification wastewater of Semen Armeniacae Amarums is adsorbed by ferrous sulfate heptahydrate, the cyanide concentration in the soaking solution can be reduced to about 2.70×10-5mol/L, which is excellent for the recycling use of the treated soaking solution in the detoxification of Semen Armeniacae Amarums soaking. After seven cycles of cyclic soaking, most of the hydrocyanic acid in Semen Armeniacae Amarums has been removed, the ferrous sulfate heptahydrate is inexpensive, the treatment is simple and effective, and it can be used as a reference for the detoxification process of Semen Armeniacae Amarums.

References

1. Li Q, Li Y. Study on acid hydrolysis condition of amygdalin[J]. Mod Chinese Med, 2009: 23-25.
2. Organization W H. WHO | Chemical hazards in drinking-water - cyanide[J]. World Health Organization, 2014.
3. Chao L I, Tong M, Pan R, et al. Summary of cyanide containing wastewater treatment[J]. Shanxi Chemical Industry, 2017,37(06):149-151.
4. Xueping L, Hong Z, Li Z. Research progress of cyanide-containing wastewater treatment[J]. Journal of Chemical Industry & Engineering, 2012, 33(2): 17-23.
5. Hendrickson T N, Daignault L G. Treatment of complex cyanide compounds for reuse or disposal[M]. US Government Printing Office, 1973.
6. Adams M D. The removal of cyanide from aqueous solution by the use of ferrous sulphate[J]. Journal of the Southern African Institute of Mining and Metallurgy, 1992, 92(1): 17-25.
7. Huda M F, Helmy Q. Removal of cyanide from alumina smelter wastewater using precipitation and filtration technique[C]//IOP Conference Series: Earth and Environmental Science. IOP Publishing, 2021, 896(1): 012075.
8. Jin X, Zhu X, Shu-Guang L V. Study on optimization of cyanide removal process for coking wastewater[J]. Environmental Pollution & Control, 2015,37(06):79-82.
9. Mishra R K. Cyanide destruction and gold recovery - a review [J]. Precious Metals, 2002(26): 44-65.
10. Beckenn W E, WE B. Treatment of cyanide waste[J]. Electroplating and Metal Finishing, 1972, 25(12): 20-21.

Electronic Properties of MTe_2/AsI_3(M=Mo and W) Van der Waals Heterostructures

Zixiang Liu And Zhiguo Wang*

School of Electronics Science and Engineering, University of Electronic Science and Technology of China, Chengdu, 610054, China

Abstract. Two dimensional (2D) materials with unique physical or chemical prperties has triggered worldwide interest in the fields of material science, condensed matter physics, and devices physics. Vertically stacking different 2D materials enables the creation of a large variety of van der Waals heterostructures. The van der Waals heterostructures robust the merits of the 2D materials electronic prperties. Using density functional theory calculations, the electronic structure of MTe_2/AsI_3(M=Mo and W) Van der Waals heterostructures are investigated in this work. The results show by stacking MTe_2(M=Mo and W) and AsI_3 vertically, a strongly binding vdW heterostructure with a type-II band alignment can be formed, which gives expectation of high multifunctional electronic performance. This theoretical study provides vital insights of 2D materials and their heterostructures which could be potential candidates for future nanoelectronic applications.

1. Introduction

Two dimensional layered materials, such as graphene, hexagonal boron nitride, transition metal oxides, transition metal dichalcogenides (TMDs), and metal halogenides have been extensively explored due to their excellent electronic, optical and catalytic properties [1, 2]. Among them, TMDs layers with direct bandgap, high stability, and versatile phases have gained tremendous attention in recent years due to their great potential in photovoltaics, hydrogen evolution catalysts, lithium-ion batteries, and the field-effect transistors (FETs) [3]. Among the TMDs, $MoTe_2$ and WTe_2 monolayers have been investigated theoretically and experimentally due to their novel electronic properties [4, 5]. Recently, a series of new layered materials such as metal trihalides MX_3 (AsI_3, ScI_3, SbI_3, YI_3, BiI_3, $ScCl_3$, $ScBr_3$ and YBr_3), have been theoretically investigated by Liu et al [2]. The MX_3 monolayers are thermodynamically stable and feasible to be exfoliated from their corresponding layered bulk materials [6]. Especially, the AsI_3 monolayer is potential photocatalyst for water splitting with indirect band gaps of 2.00eV and moderate electron mobility [1].

Over the past decades, van der Waals hetero structures have gained significant attention due to their abundant materials systems and controllable electronical properties [7]. By combining different2D materials to form van der Waals heterostructures, it is possible to overcome the drawback of individual 2D materials [8]. Vertically stacking the layer by layer arrangement of 2D materials offers an interesting way to build heterostructure multilayers in controlled manner. In this work we present the electronic properties of MTe_2/AsI_3(M=Mo and W) van der Waals heterostructures using GGA-PBE method with their pristine structures.

2. Computational methods

All calculations here are performed by density functional theory (DFT) methods using Vienna Ab initio Simulation Package (VASP) [8]. The Projector augmented wave (PAW) method is adopted to describe electron-ion interaction [9]. And Perdew-Burke-Ernzerhof (PBE) exchange-correlation functional within a generalized gradient approximation (GGA) is employed, while a kinetic-energy cutoff of 520 eV is used. The van der Waals interactions resulting from dynamical correlations between fluctuating charge distribution was described within the DFT-D2 method of Grimme [10]. The self-consistent filed (SCF) calculations are performed with an energy and force convergence of 10^{-6}eV and 0.02eV/Å, respectively. To avoid the interactions between two adjacent periodic images, the vacuum thickness was set to be 20 Å. Monkhorst-Pack grid of $10\times10\times1$ have been used to sample the brillouin-zone.

3. Results and discussion

We perform the density functional theory (DFT) calculation with the full relaxation of pristine $MoTe_2$, WTe_2 and AsI_3 monolayers. The MTe_2(M=Mo and W) monolayer consists of an interlayer hexagonal plane of Mo (W) atoms sandwiched by two hexagonal planes of Te atoms in a trigonal prismatic arrangement as shown in Fig.1 (a-b). The optimized lattice constant of 3.55 and 3.56Å for $MoTe_2$, WTe_2 monolayer, respectively. Fig.1(c)

* Corresponding author: zgwang@uestc.edu.cn

https://doi.org/10.1051/matecconf/202338001011

shows from top view of AsI_3 monolayer, each As atom is coordinated to six I atoms in an octahedral structure and from the side view there are three atom layers. The optimized lattice constant of AsI_3 monolayer is 7.29Å. These calculated values are agree well with previous theoretical report [1, 4]. Therefore, as two $MoTe_2$ (WTe_2) monolayer cells stacked on one cell of AsI_3 monolayer to construct $MoTe_2/AsI_3$ (WTe_2/AsI_3) vdW heterostructure (as shown in Fig. 1e or f for one example), the lattice mismatch between the MTe_2(M=Mo and W) and AsI_3 monolayer is only within 2.5%. The minimization of total energy results in optimized vertical interlayer separation of 3.67 and 3.65 Å for $MoTe_2/AsI_3$ and WTe_2/AsI_3 heterostructures. The structural stability of $MoTe_2/AsI_3$ and WTe_2/AsI_3 heterostructure are evaluated by calculating the binding energy (BE) which is defined as:

$$E_{binding}=E_{MoTe2/AsI3}-E_{MTe2}-E_{AsI3} \quad (1)$$

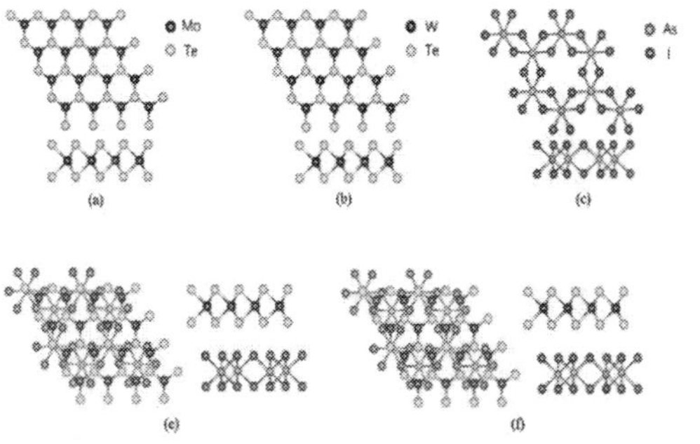

Fig. 1. Ball and stick model with respective top and side view for pristine (a)$MoTe_2$ (b) WTe_2, (c)AsI_3 monolayer and (d) $MoTe_2/AsI_3$ (e) WTe_2/AsI_3 van der Waals hetero structures.

where $E_{MTe2/AsI3}$ is the total energy of vdW hetero structures and E_{MTe2} and E_{AsI3} represent the total energies of pristine MTe_2(M=Mo and W) and AsI_3 monolayer, respectively. The binding energy is found to be -4.38 and -4.56eV with $MoTe_2/AsI_3$ and WTe_2/AsI_3 hetero structures, respectively. The negative adsorption energies indicate the exothermic process.

As shown in Fig.2(a-b), either $MoTe_2$ or WTe_2 monolayer monolayer is a direct semiconductor with the bandgap values of 1.07 eV and 1.06 eV, respectively, which are corresponding to previous study[4]. PDOS shows that the VBM and CBM of both $MoTe_2$ and WTe_2 monolayer are mainly contributed by the Mo atoms. Fig.2(c) shows the projected band structure of pristine AsI_3 monolayer which is an indirect band gap semiconductor of 2.23eV with CBM at Γ point, and VBM at between Γ and M point. Fig.2(d) shows its heterostructure with $MoTe_2$ which has been calculated along high symmetry Γ-M-K-Γ directions. The $MoTe_2/AsI_3$ vdW heterostructure is an indirect semi conductor with 1.06eV, with the CBM at M point and VBM at K point. The PDOS analysis suggests that contribution to VBM originates from Mo atom and CBM from I atoms. Fig.2(f) shows WTe_2/AsI_3 vdW heterostructure is also an indirect semiconductor with band gap of 0.91eV which CBM at M point and VBM at K point. The PDOS analysis suggests that contribution to CBM originates from I atom and VBM from W atoms. These PDOS indicate that VBM and CBM are localized on different layers suggesting the physical separation of charge carriers.

Fig.3(a) shows the electrostatic potential of $MoTe_2/AsI_3$ heterostructure along the z direction. The result shows that $MoTe_2$ has deeper potential than that of the AsI_3 with difference of 2.3eV, implying that a strong electrostatic potential exists at the interface. Fig.3(b) shows the electrostatic potential of WTe_2/AsI_3 hetero structure along the z direction. The result implies that $MoTe_2$ also has deeper potential than that of the AsI_3 with difference of 2.29eV, which indicates that the facilitated electrons transform to WTe_2 monolayer. Furthermore, the band alignments of pristine MTe_2, AsI_3 monolayers, and $MoTe_2/AsI_3$ heterostructure are illustrated in Fig. 3(c). The position of CBM and VBM bands of $MoTe_2$ is higher than that of AsI_3 ones after stacking into heterostructure and also the steps in the conduction and valence bands go in the same direction, which makes the $MoTe_2/AsI_3$ heterostructure type II band alignment. As shown in Fig.3(d), the result indicates that WTe_2/AsI_3 hetero structure is also type II band alignment.

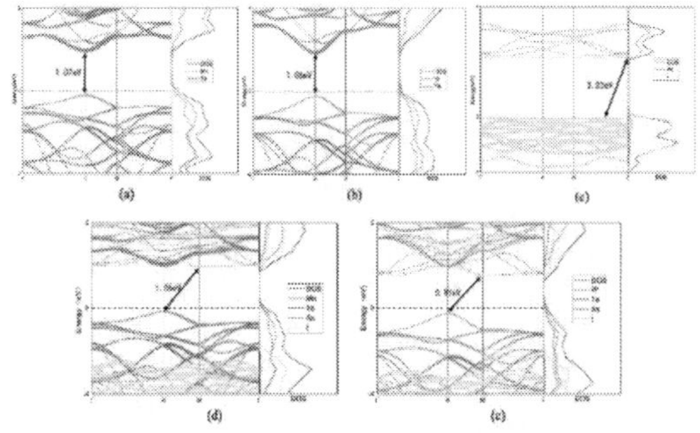

Fig. 2. Projected band structures and PDOS of pristine (a) $MoTe_2$, (b) WTe_2, (c) AsI_3 monolayers, (d) $MoTe_2/AsI_3$ and (e)WTe_2/AsI_3 van der Waals heterostructures, respectively.

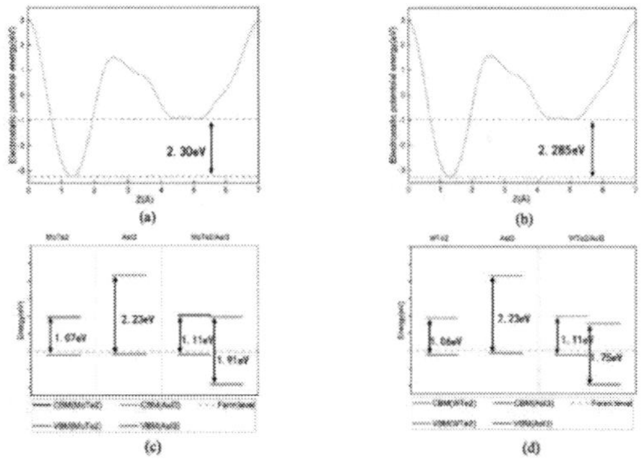

Fig. 3. (a) The electrostatic potential of $MoTe_2/AsI_3$ hetero structure along the z direction, (b) The electrostatic potential of WTe_2/AsI_3 heterostructure along the z direction, (c)band edge alignment of $MoTe_2$, AsI_3 monolayer, and $MoTe_2/AsI_3$ hetero structure (d) band edge alignment of WTe_2, AsI_3 monolayer, and WTe_2/AsI_3 heterostructure, respectively.

 https://doi.org/10.1051/matecconf/202338001011

4. Conclusion

In conclusion, the structural and electronic properties of MTe_2/AsI_3(M=Mo and W) van der Waals hetero structures have been investigated by using DFT simulation. We find a strong binding of -4.38 and -4.56eV with $MoTe_2/AsI_3$ and WTe_2/AsI_3 hetero structures, respectively. The band gap of 1.06 and 0.91 eV has been found in $MoTe_2/AsI_3$ and WTe_2/AsI_3 heterostructure, respectively. The band alignments of the MTe_2/AsI_3 (M=Mo and W) heterostructures are type II. These results reveal that it is worth investigating other heterostructure from the family of transition metal dichalcogenides with AsI_3 for new insights and technological applications.

References

1. K. Lai, C.L. Yan, L.Q. Gao, W.B. Zhang, The J. Phys. Chem. C. 122 7656-7663 (2018).
2. P. Liu, M.K. Wu, L.Y. Li, X.G. Luo, Y.H. Cheng, H. Dong, H.J. Chen, W.C. Wang, H. Liu, W.H. Wang, F. Lu, K. Cho, J. Mater. Chem. C, 5 9066-9071 (2017).
3. Y. Zhao, Y. Li, M.X. Liu, K.W. Xu, F. Ma, J. Phys. Chem. C, 124 4299-4307 (2020).
4. J. Liu, Y.Q. Ma, Y.W. Dai, Y. Chen, Y. Li, Y.N. Tang, X.Q. Dai, Chin. Phys. B, 28 107101 (2019).
5. A.N. Domozhirova, A.A. Makhnev, E.I. Shreder, S. V. Naumov, A. V.Lukoyanov, V. V.Chistyakov, J. C. A. Huang, A. A. Semiannikova, P. S. Korenistov, V. V Marchenkov, J. Phys.: Conf. Ser., 1389 012149 (2019).
6. P. Liu, M.K. Wu, L.Y. Li, X.G. Luo, Y.H. Cheng, H. Dong, H.J. Chen, W.C. Wang, H. Liu, W.H. Wang, F. Lu, K. Cho, J. Membr. Sci., 602 117786 (2020).
7. L. Liu, and T. Zhai, Info. Mat., 3 3-21 (2020).
8. H.L. Hou, C. Anichini, P. Samorì, A. Criado,and M. Prato, Adv. Funct. Mater., 32 2207065 (2022).
9. G. Kresse, J. Furthmüller, Phys. Rev. B, 54 169-186 (1996).
10. G. Kresse, J. Furthmüller, Comput. Mater. Sci., 6 15-50 (1996).
11. J.Harl, L. Schimka, and G. Kresse, Phys. Rev. B, 81 11 (2010).

https://doi.org/10.1051/matecconf/202338001012

Application of Alkyl Amidopropyl Betaine in Fire Fighting Foam Extinguishing Agent

Jiancheng Zhang*, Zeyao Shi, Meiqi Zhang and Liusuo Wu

Beijing Nari Yihe Environmental Protection Technology Co., Ltd, Beijing, China

Abstract. In order to improve the fire extinguishing performance of foam fire extinguishing agents in non-polar liquid fires, the application of alkyl amidopropyl betaine with different chain lengths in aqueous film-forming foam fire extinguishing agents was studied. The relationship between the structure of alkylamidopropyl betaine and surface tension, foaming property and foam stability was analyzed. On this basis, different foam fire extinguishing agent formulations were formed, and then the surface tension, foam performance and fire extinguishing performance of each formulation were tested. The results show that the alkyl chain of alkylamidopropyl betaine is directly proportional to the foaming property. The shorter the alkyl chain, the less oily the foam is and the better the foam's anti-burning performance. The combination of alkylamidopropyl betaine with different chain lengths is conducive to comprehensive product foam performance and oleophobic performance to achieve the best fire extinguishing effect.

In recent years, accidents of large oil-filled equipment in substations have occurred frequently, resulting in serious economic losses and social impact[1,2]. The core equipment of a substation is an oil-immersed transformer, which contains a large amount of insulating material and transformer oil.Once an oil-immersed transformer has an operating accident, it will easily lead to high-temperature transformer oil leakage and fire hazards. The national standard stipulates that "the main transformer with a single capacity not less than 125MVA should be equipped with a water spray fire extinguishing system, a synthetic foam spray system or other fixed fire extinguishing devices." Foam extinguishing agent is the most effective extinguishing medium for liquid fires, and it is suitable for preventing and controlling fires in large oil-filled equipment in substations.

At present, the performance of commercially available foam fire extinguishing agents is uneven, and the fire extinguishing performance requirements of foam fire extinguishing agents used in substations are relatively low. It is stipulated in 3.2 of " Technical standard for foam extinguishing system " [3] that when an aqueous film-forming foam liquid is used for fire extinguishing, its fire resistance level should not be lower than Class C specified in the current national standard " Foam Extinguishing Agent " [4]. When a fire occurs in a substation scene, the transformer oil burns violently. In order to ensure effective fire extinguishing of substation fires, higher fire extinguishing performance requirements should be put forward for foam fire extinguishing agents.

The key factor affecting the performance of foam fire extinguishing agent is foaming agent, which mainly includes amphoteric, nonionic, anionic and cationic surfactants. The commonly used foaming agent in the daily chemical industry is betaine amphoteric surfactant, which has good foaming ability [5] and foam viscosity. At the same time, the betaine amphoteric surfactant has good biodegradability, and the biodegradation rate can reach 81% (C14-15 betaine) and 91% (C12 betaine) in the closed bottle test [6]. Considering its low biological irritation, good resistance to hard water, and strong compounding ability, the application of betaine amphoteric surfactants in fire-fighting foam fire extinguishing agents has great prospects.

In order to improve the fire extinguishing efficiency of foam fire extinguishing agent for Class B fire combustibles, this paper studies the influence of different types of alkylamidopropyl betaines on the performance of new foam fire extinguishing agents, laying a theoretical foundation for the development of environmentally friendly and efficient aqueous film-forming foam fire extinguishing agents, but also provides technical support for the safe operation of the substation.

1. Materials and Method

1.1 Main instruments and reagents

Capstone 1470, a betaine-type fluorosurfactant with an active ingredient of about 27wt%, was purchased from Chemours Chemical (Shanghai) Co., Ltd. Alkyl glycoside APG0810, 50wt% aqueous solution, was purchased from Yangzhou Chenhua New Materials Co., Ltd. Alkyl amidopropyl betaine ODAB, OLAB, CAB, 35wt% aqueous solution, BASF (China) Co., Ltd.; xanthan gum, food grade, Zhongxuan Biochemical Co., Ltd. Diethylene glycol butyl ether, urea, sodium benzoate, domestic analytical grade.

* Corresponding author: 18810612319@163.com

G6500 quadrupole-time-of-flight tandem mass spectrometer, Agilent, USA; QBZY-1 automatic surface tensiometer, Shanghai Fangrui Instrument Co., Ltd., China; 2152 Roche Foam Instrument, Beijing Hengaode Instrument Co., Ltd., China; rotation Viscometer, Brookfield, USA; foam generating device, in line with GB15308-2006 standard. HH-501 super constant temperature water bath, Jintan Baita Xinbao Instrument Factory, China; RW20 digital overhead mechanical stirrer, IKA, Germany.

1.2 Experimental method

The relative molecular masses of the three alkylamidopropyl betaines were determined by mass spectrometer in positive ion mode.

An aqueous solution with a mass fraction of 2% surfactant was prepared, and the surface tension of the solution at 20°C was measured using a fully automatic surface tensiometer. The measurement results were obtained by taking the average value of two experiments.

The Foam test of alkylamidopropyl betaines refers to the determination method of surfactant foaming power [7]. Under the condition that the temperature of the aqueous solution is 50°C, the foam volume of 500mL surfactant aqueous solution dropped from a height of 450mm was measured.

The foam solution and deionized water were prepared at a volume ratio of 3:97; the surface tension of the foam solution at (20±1)°C was measured; (5-7) mm thick cyclohexane was added to the foam solution, The interfacial tension was measured after waiting 6 min. The measurement results were obtained by taking the average value of two experiments.

Calculate the diffusion coefficient between the foam solution and cyclohexane according to formula (1):

$$S=\gamma_c-\gamma_f-\gamma_i \tag{1}$$

In the formula: S——diffusion coefficient, mN/m;

γ_c—the surface tension of cyclohexane, mN/m;

γ_f—the surface tension of the foam solution, mN/m;

γ_i — the difference between the foam solution and cyclohexane

The foam generating device is built according to the national standard "Foam Fire Extinguishing Agent", as shown in Figure 1. First, the inlet pressure of the foam gun was adjusted to 0.63±0.03MPa, and the foam expansion ratio and 25% draining time of the foam fire extinguishing agent were tested in a 11.4L/min foam generator.

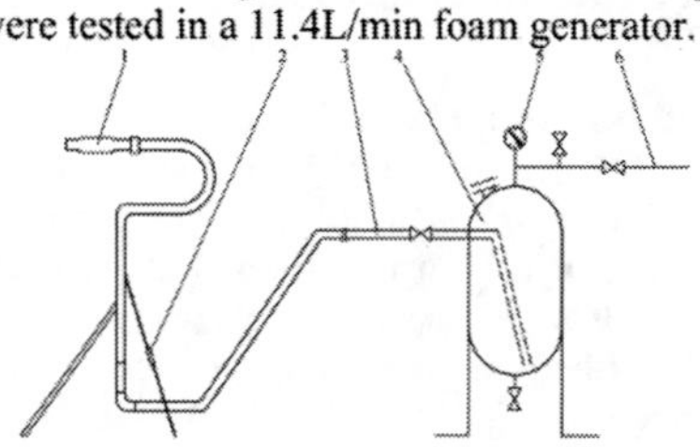

1——Standard foam gun; 2——Adjustable bracket;
3——foam delivery pipe; 4——Pressure storage tank;
5——Pressure gauge; 6——intake pipe

Fig. 1. Standard foam generating device

Referring to the fire extinguishing performance test method of low-expansion foam fire extinguishing agent, the fire extinguishing performance of different fire extinguishing agent series on rubber industrial solvent oil fires was tested. The area of the oil pan is set to 4.52m^2, the fuel consumption is set to 144L, and the preburn time is set to 60s. Adjust nozzle to ensure that the jet continues to hit the centre of the backboard, apply foam for (180±2) s.If the fire within the fire tray is extinguished prior to the end of foam application, record the extinction time as the period from the start of foam application until the time all flames are extinguished within the fire tray.

After a further (300±10)s, place a burn-back pot, containing (2±0.1)L of fuel in the centre of the tray and ignite. Record the 25 % burn-back time.

2. Results and discussion

2.1 Product structure test

The mass spectra of three different alkylamidopropyl betaines are shown in Figure 2. In the positive ion mode of mass spectrometry, the charged particles in the mobile phase are generally $[M+H]^+$, $[M+NH_4]^+$, $[M+Na]^+$, $[M+K]^+$. The difference between the molecular weight of the sample and the characteristic value determined by mass spectrometry is 23, which is consistent with the structural characteristics of [M+Na]+. The main structure of CAB is cocamidopropyl betaine, and the carbon-carbon alkyl chain length is between 12 and 18. The carbon-carbon alkyl chain length of ODAB and OLAB is between 8 and 12. The main structure of ODAB is decyl amidopropyl betaine, and the main structure of OLAB is dodecyl amidopropyl betaine.

m/z(ODAB)= 309.07, 337.10, 365.08

m/z(OLAB)=308.96, 337.04, 365.09

m/z(CAB)=365.09, 393.20, 421.24, 449.18

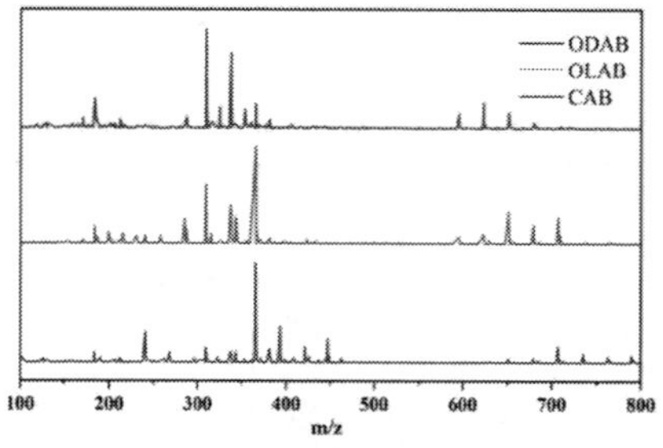

Fig. 2. Mass spectra of ODAB, OLAB, CAB

2.2 Surface activity test

Surface activity is usually expressed by critical micelle concentration cmc and critical surface tension γcmc. They respectively represent the ability of the system to form micelles and the ability to reduce the surface tension of the liquid [8]. Surface tension tests of alkylamidopropyl betaines with different alkyl chain lengths are shown in Figure 3. The surface tension of alkylamidopropyl betaine decreased first and then increased with the increase of mass concentration. In contrast, CAB had the lowest critical micelle concentration. When the mass concentration is 0.1%, the surface tension reaches 34.0 mN/m. When the amount of OLAB added was 0.2%, the surface tension reached a minimum of 29.5 mN/m. When the addition of ODAB is 0.5%, the surface tension reaches the minimum of 27.3mN/m. Alkyl chain shortening of surfactants leads to an increase in critical micelle concentration and a decrease in critical surface tension.

The surface activity of surfactants is affected by the chain length of the hydrophobic group and the chain length distribution. The cmc value mainly depends on the length of the hydrophobic group of the surfactant. The larger the chain length of the alkyl chain, the smaller the cmc.γcmc is less affected by the chain length of the hydrophobic group of the surfactant, but greatly affected by the chain length distribution of the surfactant molecule in the saturated adsorption layer. The closer the chain length distribution, the smaller the γcmc value.

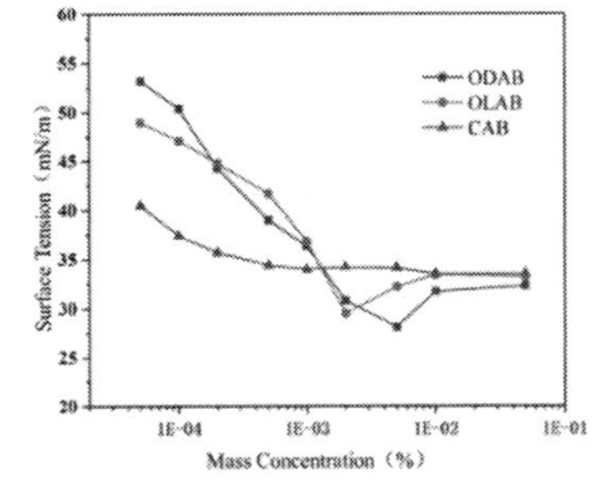

Fig.3. Surface tension of different types of alkylamidopropyl betaine

2.3 Foaming ability and foam stability test

The foamability and foam stability of different alkylamidopropyl betaine aqueous solutions were tested. As shown in Figure 4, the increase of the alkyl chain greatly improves the foamability and foam stability of the surfactant. In general, the lower the critical micelle concentration of a surfactant, the stronger the ability to generate foam. The critical micelle concentration of CAB is lower than that of OLAB and ODAB, so CAB shows more excellent foaming ability at the same concentration.

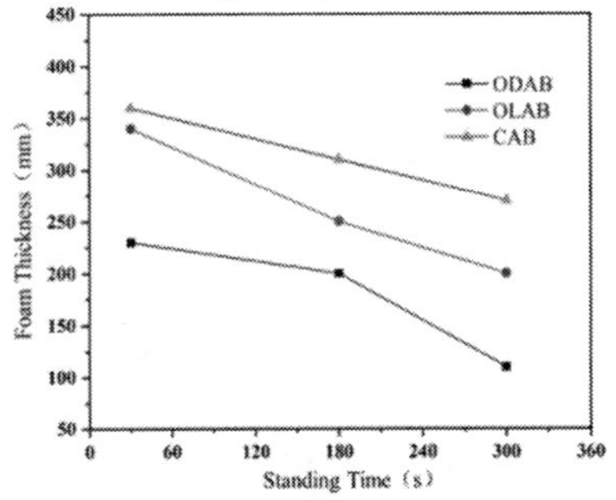

Fig.4. Foam characteristics of different types of alkylamidopropyl betaine

The structure of the surfactant also has a certain influence on the stability of the foam. The length of the hydrophobic chain can affect the arrangement of the surfactant in the liquid film and the diffusion rate of the gas through the liquid film, thereby affecting the life of the foam.

As shown in Figure 5, the surfactant is adsorbed on the air-liquid interface of the bubbles. The hydrophilic polar chain ends extend into the water, and the hydrophobic hydrocarbon chains extend in the gas phase. The surfactant aligns and forms bimolecular film bubbles. Bubbles can achieve long lifetimes under the protection of bilayer membranes. As the bubbles are continuously generated, they gradually accumulate on the surface of the liquid to form a foam layer. The growth of the hydrophobic chain to a certain extent is conducive to increasing the interaction force between surfactant molecules, making the arrangement of non-polar groups more compact, thereby enhancing the foam stability.

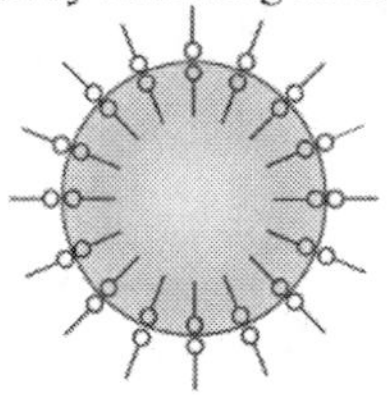

Fig. 5. The molecular arrangement of surfactants in the foam

2.4 Preparation of foam fire extinguishing agents with different formulas

The existing foam fire extinguishing agent formula is composed of Capstone 1470, APG0810, CAB, ethylene glycol, diethylene glycol butyl ether, urea, xanthan gum and water. Based on the existing product formula, ODAB and OLAB are used to replace CAB in the foam fire extinguishing agent. The main components and the addition amount are shown in Table 1. The formed foam fire extinguishing agent is mixed with water at a volume ratio of 3:97.

Table 1. Setting parameters of foaming agent in foam fire extinguishing agent formulation

	APG	ODAB	OLAB	CAB
Sample 1	3.0	—	—	12.0
Sample 2	3.0	—	12.0	—
Sample 3	3.0	12.0	—	—
Sample 4	3.0	—	8.5	3.5
Sample 5	3.0	8.5	—	3.5

2.5 Physicochemical performance analysis of formula

The surface tension, interfacial tension, foaming ratio, and 25% draining time of different product formulations were tested. Surface tension, interfacial tension, and diffusion coefficient are the main ways to characterize the spreading properties of foam liquid. As shown in Table 2, the use of fluorosurfactants in combination with hydrocarbon surfactants exhibits lower surface tension (<18mN/m). When the hydrocarbon surfactant is replaced with a shorter alkyl chain betaine surfactant, the formulation has lower surface tension. Capstone 1470 is a perfluoroalkyl betaine with a relatively short fluoroalkyl chain (C6), which has good environmental protection and low toxicity[9]. As the length of the hydrophobic chain of the alkyl betaine surfactant increases, the hydrocarbon chain may bend to cover part of the fluorocarbon chain. This phenomenon causes a relative decrease in the density of the outermost fluorocarbon chains, resulting in an increase in surface tension.

Table 2. Physical and chemical performance test parameters of different formulations

	Surface Tension (mN/m)	interfacial tension (mN/m)	diffusion coefficient	Foaming rate (times)	drainage time (min)
Sample 1	17.2	2.2	5.6	7.9	3.8
Sample 2	16.9	2.2	5.9	7.0	3.7
Sample 3	16.8	2.2	6.0	5.8	3.2
Sample 4	17.1	2.2	5.7	7.5	4.2
Sample 5	16.8	2.2	6.0	7.4	5.0

It can be seen from Table 2 that the increase of the hydrophobic chain length of alkylamidopropyl betaine significantly improves the foaming ability and 25% draining time of the formula. This is consistent with the foaming and foam stabilizing properties of surfactants. Adding a small amount of CAB on the basis of ODAB/OLAB can also improve the foam characteristics of the product and achieve better foaming ability and foam stability.

2.6 Analysis of formula fire extinguishing performance

Different formulations have completed the performance test of strong release fire extinguishing. As shown in Table 3, when CAB is used as the main blowing agent, the ability to control the fire at the edge of the oil pan is low, the sealing ability of the foam and water film is poor, the flame can burn through the foam, and the fire cannot be extinguished within 3 minutes. When the short alkyl chain betaine surfactant is used as the main foaming agent, it has a strong ability to control and extinguish the fire of fuel. When OLAB is used as the main foaming agent, the product has strong fire extinguishing ability. When ODAB is used as the main foaming agent, the foam's anti-reburning ability is greatly improved.

Table 3. Fire extinguishing and anti-burn performance parameters of different formulations

	Extinguishing time /min	burn-back time /min
Sample1	>3	6.3
Sample2	2.0	12.3
Sample 3	2.2	16.2
Sample 4	1.9	12.5
Sample 5	2.6	11.1

The longer the non-polar alkyl chain of the surfactant, the better the lipophilicity of the foam produced, and the worse the blocking ability of the fuel. Therefore, when CAB is used as the main foaming agent, the fire extinguishing performance of the product is greatly reduced. The shortening of the alkyl chain increases the HLB value, and the hydrophilic ability of the formed foam is enhanced, which has a stronger sealing ability for liquid combustibles, so the fire extinguishing effect is greatly enhanced. When ODAB is used as the main blowing agent, the foam performance of the product is relatively poor, and the fire extinguishing ability is reduced. As shown in formulation 4, the combination of betaine surfactants with different alkyl chains can improve the foam coverage and water film sealing characteristics of the product. On the one hand, a small amount of long alkyl chain betaine surfactant can greatly improve the foaming ability and foam stability of the foam. On the other hand, a large amount of short alkyl chain betaine surfactants can improve the problem of poor foam sealing. Under the double action of foam and water film, the best effect of fire extinguishing and anti-resurgence is achieved.

3. Conclusion

The article studies the application of different chain lengths of alkylamidopropyl betaines in aqueous film-forming foam fire extinguishing agents, and draws the following conclusions:

(1) The longer the hydrophobic chain segment of alkylamidopropyl betaine, the higher the surface activity and the better the foam performance.

(2) The structure of alkylamidopropyl betaine has little influence on the spreading performance of the foam fire extinguishing agent system, but has a greater influence on the foaming performance and foam stability of the product. The hydrophobic chain growth of alkylamidopropyl betaine is beneficial to Improve the foam performance of the product.

(3) The shortening of the hydrophobic chain length of alkylamidopropyl betaine is conducive to the improvement of fire extinguishing performance. The combination of betaine surfactants with different alkyl chains can improve the foam coverage and water film sealing characteristics of the product.

References

1. YL Wang, CH Li, JQ Zhang, FJ Zhang, SX Lu, MH Fan, LF Wang, J. Saf. Environ 26, (6) 2019.
2. MH Li, W Li, XF Du, J Qi, SP Wang, HC Wu, East China Electric Power 41, (9)2013.
3. Technical standard for foam extinguishing system, GB 50151-2019.
4. Foam Extinguishing Agent, GB15308-2006.
5. SD Niu, JK Du, Journal of China People's Police University 30, (2)2014.
6. YQ Ren, XC Wang, TT Qiang, Leather and Chemicals 26, (1)2019.
7. Surface active agents--Measurement of foaming power--Modified Ross-Miles method, GB/T 7462-1994.
8. BY Zhu, GX Zhao, Fine Chemicals 4,1987.

Research on the optimization of electrochemical behavior model in the environment of novel composite fuel cell

Ling Chen*, Caibao Han, Chunxia Wang and Ting Zhu

School of elementary education, Wanjiang University of Technology, Ma'anshan 243000, China

Abstract. Composite materials have been widely studied due to their good mechanical, electrical and chemical properties, especially atomic structure fuel cells with large specific surface area, high electrical conductivity and good compatibility of composite materials. These new composite materials have many excellent physical and chemical properties, and are widely used in many fields, especially in the field of electrochemistry, showing their unique advantages. Composite materials play an important role in the electrocatalytic reaction of composite materials due to their high specific surface area and good compatibility of composite materials. The construction based on composite materials has high sensitivity, wide linear range, good repeatability and stability. As a catalyst in a fuel cell, the composite material can improve the energy density and fuel utilization rate of the fuel cell, which is a high-performance fuel cell.

Keywords: Composite material; fuel cell; electrochemical behavior model.

1. Introduction

The fuel cell is a power generation device that directly converts the chemical energy in the fuel and the oxidant into electrical energy, and the proton conductor solid oxide fuel cell has the characteristics of not generating water vapor at the anode and the electrolyte has a lower activation energy for conducting ions. Compared with Other fuel cells have higher equilibrium electromotive force and work characteristics at medium temperature . Its wide applicability to fuel gas, high power generation efficiency, high energy density, high value of waste heat utilization, all-solid structure and easy assembly have become one of the focuses of researchers [1]. During the working process of the fuel cell, the mutual conversion process of chemical energy, electric energy and thermal energy is extremely complicated. In order to obtain the highest possible battery output power and less thermal stress, in order to improve the performance of the battery and prolong its service life, it is necessary to Modeling research on batteries is conducive to understanding the electrochemical reaction process and material migration process inside the battery, and plays an important guiding role in experimental research and improving battery performance [2].

2. The concept of composite materials

Composite material is a multidisciplinary emerging discipline developed in the early 1990s. Its development has opened up a new level of human understanding of the world. From the perspective of the structural unit level of the material, it is in the mesoscopic field[3]. It has four outstanding effects, such as surface effect, volume effect, quantum effect and macroscopic quantum tunneling effect. It shows unique characteristics in the fields of optoelectronic devices and catalysis. Composite materials are widely studied and applied materials. A composite material refers to a composite material in which the scale of the dispersed phase has at least one dimension smaller than 100 nm. The dispersed phase can be composed of atoms or non-atoms [4]. The composite material studied in this paper is mainly a material in which atoms are closely packed into a single-layer two-dimensional honeycomb lattice structure, which can be regarded as the basic unit for constructing other dimensional materials [5]. It has the characteristics of high electrical conductivity, high mechanical properties, and super large specific surface area, and is widely used in the fields of optoelectronics, new energy, and catalysis. It is hollow formed by single-layer or multi-layer graphite sheets wound around the central axis at a certain helical angle, and each layer is a cylinder composed of a hexagon formed by one atom hybridized and fully bonded to the surrounding three atoms. noodle. According to different formation conditions, atomic structure nanotubes are divided into multi-wall atomic structure nanotubes and single-wall atomic structure[6]. Atomic structure nanotubes have good electrical conductivity, mechanical properties and compatibility of composite materials, and can be used in optoelectronics, energy storage devices, medicine and other fields. It is a fibrous nanocomposite material curled from multilayer graphite sheets, with a

* Corresponding author: chenling@hhuwtian.edu.cn

diameter ranging from tens to 200nm and a length in the micron range. It is a one-dimensional composite material between ordinary atomic structure and atomic structure nanotubes. According to the positional relationship between different nanolayers and the fiber axis, it can be divided into three types: tubular fibers, sheet fibers, and herringbone fibers. High molecular polymers have very stable chemical properties and high corrosion resistance, and can be used as polymer reinforcements, catalyst supports, and gas storage systems.See Figure 1.

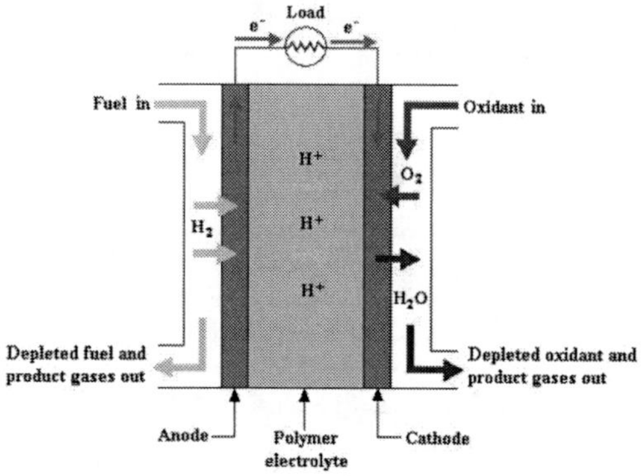

Fig. 1 Application Principles of Composite Materials in Fuel Cells

3. Application principles of composite materials in fuel cells

Fuel cells generate electric current through the chemical reaction between fuel and oxidants on electrodes, and are important devices for energy transfer and storage. Under unique oxidation conditions, a part of the outer wall of the atomic structure nanotube is opened, and the nanoscale graphene layer attached to the inner wall is exposed, and the inner wall of the nanotube is still intact and conductive, which promotes the electrocatalytic process. The charge transfer, thus promoting the formation of active sites of the catalyst, the composite improves the catalytic activity of the catalyst.

3.1 Application in the field of fuel cell catalysts

Although composite materials have been widely used in proton exchange membrane fuel cells, they do not have catalytic activity for oxygen reduction, which greatly reduces the efficiency of use. The study found that in the active sites around the atoms, the nitrogen atoms with high electronegativity will take away the electrons of the atoms, and the atoms that lose electrons will promote the activation of oxygen molecules. Since the molecules of the composite material contain nitrogen atoms, it can be used as a nitrogen source for doped graphene, and the doped graphene is used as an electrode catalyst for a proton exchange membrane fuel cell. , the electron transfer number of oxygen reduction reaction formed by calcination at 600 ℃ is 3.8, indicating that it has super stability. In addition, the dynamic limit current density of the 9mA/cm^2, the kinetic limit current density of 8.5mA/cm^2, because at 600 ℃ , the decomposition reaction of the composite material, the loss of nitrogen atom functionality, the catalytic activity decreased.

Through the comparison of electrochemical properties, it is found that the composite material has a cathode peak at -0.44V, and the peak value is -0.5mA/cm^2, and the composite material has a cathode peak at -0.45V, and the peak value It is -1.5mA/cm^2, indicating that the composite material has a better adsorption effect, and the content of oxygen molecules on the surface decreases. In addition, the number of electrons transferred per unit oxygen molecule of the composite material is 0.72, 0.76, 0.13 when the electrode potential is -0.6V, -0.7V, -0.8V. It shows that the composite material has high nitrogen content. By integrating nanowires and graphene to form a unique nanostructure, it can promote the effective adsorption of oxygen molecules, improve the efficiency of electron transfer, and increase the catalytic activity for oxygen reduction reaction. The application process of composite fuel cell catalysts is shown in Fig. 2 (a) and (b).

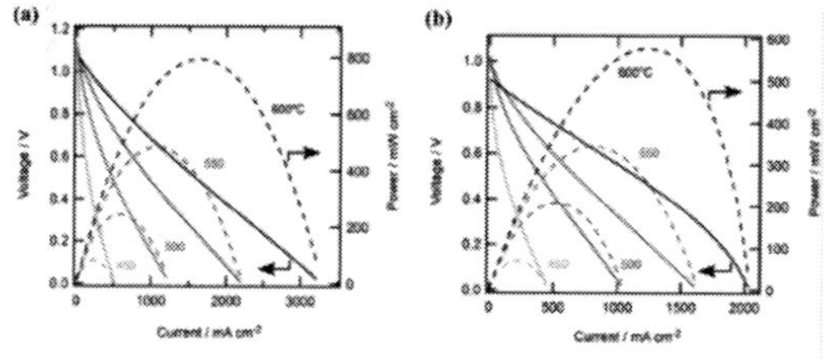

Fig. 2 Application Process of Composite Fuel Cell Catalyst

3.2 Electrochemical performance characterization comparison

The results show that the current density of the composite material is close to 0 when the electrode potential is -0.8V, and the current density of the composite material is -2.0mA/cm^2. In addition, the number of electrons transferred per oxygen molecule of the composite material calcined at 600°C was 0.72, 0.78, and 0.86 at electrode potentials of -0.4V, -0.5V, and -0.6V, respectively. It shows that the composite material can be used as an efficient catalyst for the oxygen reduction reaction and be used in scientific research. The electrochemical impedance method test found that the electron transfer resistance of the composite material was 356.2Ω and 359.97Ω respectively, and the Rct of the composite material was 331Ω. Compared with the general anode catalyst, the composite material showed a higher current density. and greater stability.

4. Results and discussion

4.1 Verification of calculation model

In order to verify the rationality of the model, this paper adopts the same experimental parameters as in this paper, and it is verified in the previous work. It can be seen from the current density-voltage curve that the error between the numerical calculation results and the experimental data in this paper is kept within 6.3%, which is within a reasonable range. See Figure 3.

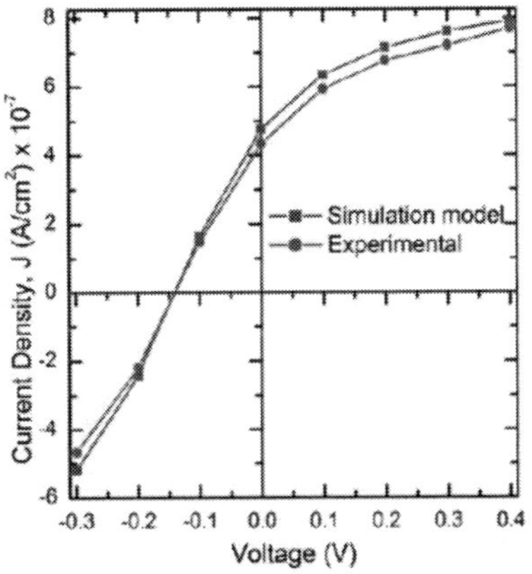

Fig. 3 Current density-voltage curve experimental simulation comparison value

4.2 Simulation results and discussion

When the fuel and air are fed in the same direction, the temperature will gradually increase along the direction of gas flow. When the reverse air intake is adopted, the temperature of the anode side increases gradually from the inlet of the fuel gas, reaches the highest value in a short distance, maintains the highest value in the middle of the battery, and then drops after a certain distance. Near the inlet of the air channel, due to the relatively low temperature of the freshly entered air, the heat generated by the electrochemical reaction is quickly absorbed and the temperature of the anode side drops; meanwhile, the temperature of the air on the cathode side rises sharply. Compared with the reverse air intake method, the temperature changes on both sides of the cathode and anode are more uniform when the air intake method is adopted in the same direction. The temperature difference between the two sides of the pole is smaller, the temperature gradient is smaller, and the thermal stress of the single cell is correspondingly reduced.

4.3 Electrochemical reaction process

As the electrochemical reaction proceeds, the temperature of the battery rises, the hydrogen is sufficient, and the current density increases. With the consumption of hydrogen, the reaction rate decreases and the current density gradually decreases. At the same time, at the cross-section at the same X position, the current density distribution in the single cell is uneven, the minimum value of the current density appears in the electrode in the middle of the flow channel on the anode side, and the maximum value of the current density appears in the current collector on the cathode side. The contact point between the plate and the cathode. Temperature distribution and power density changes at the battery reaction interface. The increase of the air intake rate provides the battery with a sufficient amount of reaction gas, and at the same time takes away a large amount of heat generated by the reaction, which reduces the gradient of the temperature field and the temperature difference between the cathode and anode plates, and the highest temperature of the battery decreases from 1186K to 1162K and At the same time, the power density of the battery increases with the increase of the flow rate. After reaching the reaction limit, the power density tends to be flat and no longer changes with the change of the air velocity. Combined with the distribution of hydrogen components, it can be known that when the intake mole fraction of hydrogen is increased, the intensity of the chemical reaction increases, the average temperature of the battery increases, the temperature difference between the internal components of the battery increases, and the temperature gradient increases, which is more likely to cause thermal stress. The power density increases proportionally. The change of hydrogen mole fraction at different positions in the anode channel. The increase of the air intake rate on the cathode side improves the fuel utilization rate on the anode side to a certain extent; at the same time, although the reaction rate can be increased, the fuel utilization rate does not increase significantly on the premise that the air supply on the cathode side is relatively sufficient.

4.4 Experiment analysis

Through the analysis of the output voltage, power and efficiency curve of the fuel cell, it can be known that the increase of the load current will reduce the output voltage, increase the output power and reduce the efficiency. On the contrary, when the load current decreases, the voltage will increase, the output power will decrease, and the efficiency will increase. The main reason is that when the load current increases, the energy consumed by the internal resistance increases and turns into heat energy, and the efficiency of the battery will decrease. When the load current decreases, the internal resistance consumption becomes smaller, thereby improving the efficiency of the battery.

5. Conclusion

Further research and exploration from the perspective of electrochemical reactions and materials is particularly necessary. By improving the microstructure of the anode, the performance of the battery can be significantly improved and the fuel utilization rate can be improved without reducing the stability of the material under reducing atmosphere conditions. With the development of numerical simulation technology, the modeling of the reaction mechanism of the battery can make up for the experimental Insufficient, provide new ideas for its performance improvement. This paper comprehensively describes the empirical model, electrochemical model and voltage dynamic model of the fuel cell, and simulates the three models through simulation tools to verify the validity and feasibility of the model, compares the differences between the models, and analyzes each characteristics of a model.

Acknowledgments

This work was financially supported by Research Fund of Wanjiang University of Technology (No: zl202212).

References

1. Jiang Dahua, Sun Kangtai, Qi Wei, etc. Current Situation and Suggestions of China's Quality Power Generation Industry. Renewable Energy, Vol. 12(2020) No.14, p.542-546.
2. Zhao Zhan, Qi Xiulong, Liu Xiaodong, et al. Performance Analysis of Mass Gasification-Fuel Cell Power Generation System. Renewable Energy, Vol. 16(2019) No.15, p.115-120.
3. Yang Siqi, Liu Zhongliang, Hou Junxian, et al. Preparation and Performance of MnO2 and Activated Carbon Mixed Catalyst for Microbial Fuel Batteries. Renewable Energy, Vol. 20(2019) No.20, p.311-316.
4. Liu Tingjie, Hu Naitao, Li Juntao, et al. Combustion pollutants of quality pellet fuel, Vol. 08(2020) No.12, p.161-163.
5. Xu Lamei, Xiao Jinsheng. Modeling and Simulation of PEM Fuel Cell Based on Electrochemical Model [. Beijing: Chemical Industry Press, Vol. 23(2020) No.16, p.71-75.
6. Yu Jiangfeng, Luo Linghong, Cheng Liang, et al. Progress in the development of materials of solid oxide fuel cells. Journal of Ceramics, Vol. 20(2020) No.19, p.613-626.

Study on Ternary Blend Organic Solar Cells Based on Multiple Non-fullerene Acceptors

Meng Wang

Department of Chemistry, Faculty of Science, Yanbian University, Yanji, Jilin, 133002, China

Abstract. In recent years, non-fullerene acceptor materials for organic solar cells (OSCs) have attracted much attention. Among them, ITIC, Y6 and perylene diimide (PDI) have been widely studied because of their easy structure regulation. The application and development of ITIC and its derivatives, Y6, PDI and its derivatives in ternary blend organic solar cells are reviewed in detail. This work focuses on the molecular structure optimization method and the selection of third component materials. At present, the efficiency of organic solar cells has exceeded 18%, but the relationship between molecular structure and photovoltaic performance needs further exploration. Optimizing the synthesis route is still an important research direction in this field.

Keywords: organic solar cells; non-fullerene receptors; ITIC; Y6; perylene diimide.

1. Introduction

In the past few decades, traditional energy sources have been unable to fully burn organic matter due to their low utilization efficiency. Therefore, too many toxic and harmful gases have been discharged into the atmosphere, and the problem of environmental pollution has gradually become prominent[1].In order to improve environmental problems and accelerate energy transformation, China has vigorously developed various new energy sources. In this context, research on organic solar cells (OSCs) is also active. At present, the non-fullerene acceptor materials in the active layer of OSCs can be divided into two types: small molecule acceptors (A-D-A, A-DA'D-A) and polymer acceptors. Many scholars have also studied the above aspects and tried to adopt different methods to improve the efficiency of OSCs.In 1958, Kearns' group at the University of California, Berkeley first discovered the photovoltaic effect of organic crystals[2]. Since then, research on OSCs has been carried out. At first, the fullerene system was the mainstream acceptor material of OSCs[3].In 2015, Zhan 's research group[4] used PTB7-TH as the donor to blend with the A-D-A type non-fullerene acceptor material ITIC, which was designed and developed for the first time. The device energy conversion efficiency (PCE) reached the highest value of 6.80% at that time, opening the non-fullerene era. In 2019, Zou 's group of Central South University[5] designed an A-DA'D-A type non-fullerene acceptor molecule Y6, which not only effectively solved the problem of mutual limitation of short-circuit current and open-circuit voltage performance, but also broke the record of PCE of single OSCs device, reaching 15.70 %. While the efficiency of organic batteries based on small molecule receptors continues to break through, polymers, especially perylene diimide (PDI) and its derivative receptors, are also widely used.

Based on this, this paper summarizes the research on ternary blend organic solar cells based on ITIC and its derivatives, Y6, perylene diimide (PDI) and its derivatives in recent years, in order to provide reference for future research on OSCs.

2. Ternary Blend Organic Solar Cells Based On A-D-A Acceptor Material Itic And Its Derivatives

In the A-D-A type acceptor material, A represents the electron-withdrawing unit and D represents the electron-donating unit. The two units are connected by the conjugated spacer unit π. This structure adopts a conjugated push-pull design, which can not only promote the transfer of intramolecular charges, but also achieve the function of each unit independently through fine modification and regulation, and improve the morphology of the active layer blend membrane. Figure 1 is the molecular formula of ITIC and its derivatives, and the corresponding device photovoltaic performance parameters are shown in table 1.

In 2015, Zhan 's group[4] synthesized the first A-D-A type non-fullerene electron acceptor ITIC based on indeno-dithienothiophene (Figure 1), where IT is the indeno-thiophene core part in the middle of the molecule, IC is the molecular end position, each end is connected with a carbonyl group and two cyano groups. The researchers combined the donor materials PTB7-Th and ITIC in different proportions and found that when PTB7-Th:ITIC=1:1.3, the battery efficiency reached a maximum of 6.80% at that time, in which the open circuit voltage (VOC) was 0.81 V, the short circuit current density (JSC) was 14.21 mA / cm^2, and the fill factor (FF) was 0.591. The synthesis of ITIC receptor materials has opened up a new research direction for OSCs, and the efficiency of OSCs based on ITIC and its derivatives has been continuously improved. This section will be divided into two parts : main side chain regulation and end group regulation.

2.1 Main-side Chain Regulation of ITIC

Small molecule receptors can regulate molecular configuration, enhance intramolecular interaction, affect intermolecular stacking behavior, enhance molecular absorption, improve molecular energy level and determine the solubility and crystallization ability of compounds through the regulation of main and side chains. In 2016, Li 's group [6] designed and synthesized a new small molecule receptor m-ITIC by changing the position of the substituted alkyl group on the side chain of ITIC, and combined it with the polymer J61 (Figure 1) [7] synthesized by the alkylation of benzenedithiophene unit. The prepared binary component J61:m-ITIC obtained 11.77 % PCE. In December of the same year, the subject[8] introduced a trialkylsilyl substituent on the dithienylbenzenedithiophene to synthesize J71 (Figure 1) with a further reduced HOMO level, and the device efficiency after blending with ITIC exceeded 11 %. In 2017, they [9] synthesized a new copolymer J91 by fluorination of side chain thiophene (Figure 1). It was found that after blending with m-ITIC, J91 reduced the carrier recombination in the blend film and inhibited the formation of triplets, and the device PCE reached 11.63 %. In 2018, Martin Heeney 's group [10] also modified the alkylbenzene side chain of ITIC and synthesized C8-ITIC (Figure 1) by alkylation pathway (first synthesized soluble and easily purified intermediates). After blending with polymer PFBDB-T, 13.20 % device efficiency was achieved. In 2020, Zhan 's group [11] synthesized the IDIC small molecule receptor based on the molecular structure of ITIC (Figure 1). The π-π stacking of the IDIC molecular end group promotes the electron migration in the active layer. The prepared FTAZ:IDIC:INIC3 three-component obtained 12.49 % efficiency.

2.2 Terminal Regulation of ITIC

The regulation of terminal groups, especially halogenation strategy, is one of the most commonly used methods in molecular modification. The regulation of small molecule receptor terminal groups can enhance intramolecular charge transfer, thereby reducing band gap, adjusting energy level and improving battery efficiency. In 2018, the Alex team[12] synthesized a small molecule 6TIC-4F with a strong power supply core unit and a strong electron-withdrawing acceptor unit on both sides based on ITIC. The photovoltaic device PCE based on PBDB-T : 6TIC-4F can reach 11.14 %. Studies have shown that such acceptor molecules can not only expand the absorption range of the active layer material to the near-infrared region, but also make the device more effectively absorb solar photons to improve battery efficiency. Hou 's group[13] synthesized a new type of small molecule acceptor IT-4F (Figure 1) by fluorination of the phenyl hydrogen atom located on the ITIC terminal group. It has higher absorption coefficient and more redshift absorption. After combining with the same fluorinated PBDB-T-SF to prepare PBDB-T-SF : IT-4F device, the PCE can reach 13.00 %. In 2019, Shen 's group[14] synthesized a small molecule receptor IN-4F with a silicon-based side chain based on IT-4F molecule. Compared with the former, IN-4F has a higher LUMO energy level and a lower optical band gap. When IN-4F is blended with PBDB-T-2F (PM6) : IT-4F, the PCE of the photovoltaic device reaches 14.90 %. In 2020, He 's group[15] introduced a single bromine atom into the IC end group to synthesize BTIC-2Br-m (Figure 1), which makes the device have a higher open circuit voltage. The introduction of different numbers of bromine atoms also makes the dipole moment significantly different, which indicates that the bromine atom plays an important role in the polarization of the molecular skeleton. The PBDB-TF : BTIC-2Br-m device based on it exhibits 16.11 % PCE. In 2021, Wang 's group[16] designed and synthesized an asymmetric halide IDTT-Cl-2F (Figure 1) with IDTT as the core unit. Based on the device prepared by blending it with polymer PBDB-T-2Cl (PM7), the efficiency of 13.30 % was obtained. Studies have shown that asymmetry and halogenation strategies have a synergistic effect on the efficiency of non-fullerene small molecule devices.

2.3 Summary

In the past two decades, the A-D-A type small molecule receptor ITIC has attracted much attention due to the advantages of easy modification and synthesis of each structural unit in the molecule, and has become one of the most potential receptors to replace fullerene. The main side chain and terminal group regulation of ITIC molecules are analyzed in detail. It is concluded that fluorination, chlorination and bromination regulation can reduce energy loss and maintain good efficiency of the device. Fluorination effectively improves the thermal stability of the acceptor molecule and increases the open circuit voltage of the device. Chlorination is more effective in reducing molecular energy levels and expanding absorption spectra. Bromination is more conducive to the transfer of charge in the device.

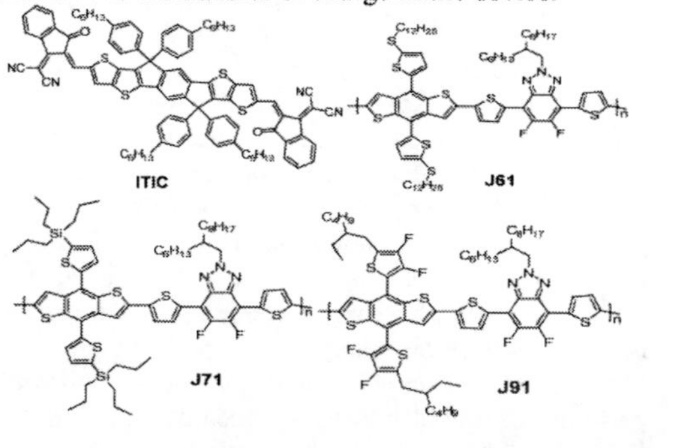

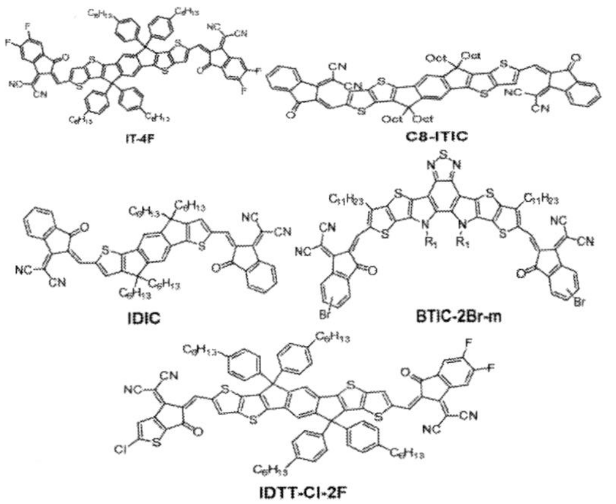

Figure 1. Molecular formula of ITIC and its derivatives

Table 1. Photovoltaic performance of ITIC and its derivatives ternary blend organic solar cell devices

Active layer	V_{OC}(V)	J_{SC}(mA/cm^2)	FF	PCE(%)	ref
PTB7-Th:ITIC	0.81	14.21	0.591	6.80	4
J61:m-ITIC	0.91	18.31	0.706	11.77	6
J71:ITIC	0.94	17.32	0.698	11.41	8
J91:m-ITIC	0.98	18.03	0.655	11.63	9
Active layer	V_{OC}(V)	J_{SC}(mA/cm^2)	FF	PCE(%)	ref
PFBDB-T:C8-ITIC	0.94	19.60	0.720	13.20	10
FTAZ:IDIC:INIC3	0.86	20.22	0.717	12.49	11
PBDB-T:6TIC-4F	0.72	23.00	0.670	11.14	12
PM6:IN-4F:IT-4F	0.86	22.80	0.762	14.90	14
PBDB-TF:BTIC-2Br-m	0.88	25.03	0.731	16.11	15
PM7:IDTT-Cl-2F	0.96	19.20	0.711	13.30	16

3. Ternary Blend Organic Solar Cells Based On A-Da'd-A Acceptor Material Y6

Based on the A-D-A type structure, Zou 's group first designed the DA 'D type fused ring skeleton by introducing chromophores and chromophores into the central fused ring, and constructed the A-DA 'D-A type receptor material. On the basis of the benzothiadiazole central unit, Y6 was synthesized by introducing multiple nitrogen-containing electron-deficient nuclear alkyl chains at the β-position of dithiophene. The emergence of Y6 increases the application range of polymer donors, fullerenes and their derivatives acceptors, non-fullerene small molecules and other substances, improves the photovoltaic performance of the battery, and promotes the development of OSCs. This section will be divided into five parts.

3.1 Polymer Donor as the Third Component

The flexibility, machinability, aggregation stability, better film quality and better photovoltaic performance of the polymer donor itself[17] have attracted the attention of researchers and continue to try to add it to the active layer to improve the efficiency of organic photovoltaic devices. Figure 2 is the molecular formula of Y6 and the polymer donor material blended with it. Table 2 is the photovoltaic performance parameters of the device corresponding to the polymer donor material. In 2019, Zou 's group[5] constructed a binary system PM6 : Y6 blended with polymer PM6 (Figure 2) based on the receptor molecule Y6 (Figure 2) they designed and synthesized. The battery efficiency exceeded 15 %, which was the world 's highest record reported at that time. In 2020, Liu 's group[18] used PM7 instead of PM6 to prepare PM7 : Y6 binary devices. PM7 has a deeper HOMO level, which helps to achieve high voltage values. After mixing the two components at 1 : 1.2, the device battery efficiency is as high as 17.04 %. Xie and co-workers[19] introduced J71[8] on the basis of PM6 : Y6. The complementary absorption spectrum of PM6 and J71 can effectively improve the photon capture rate, optimize the molecular arrangement and improve the JSC, obtaining a battery efficiency of 16.50 %. Sun 's group[20] added a wide band gap P1 (Figure 2) to PM6 : Y6. P1 has a high HOMO energy level and good compatibility with PM6. Its addition effectively reduces the non-radiative energy loss and enhances the component tolerance of the device. The battery efficiency is 16.21 %.

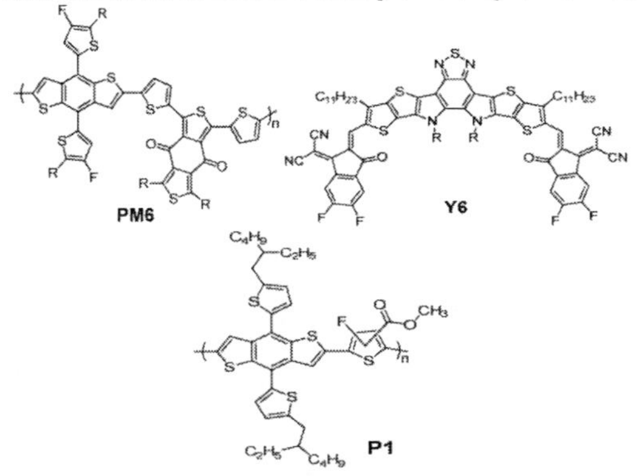

Figure 2. Y6 and molecular formula of polymer donor **materials**

Table 2. Photovoltaic performance of ternary organic solar cell devices with different polymer donor materials blended with Y6

Active layer	V_{OC}(V)	J_{SC}(mA/cm^2)	FF	PCE(%)	ref
PM6:Y6	0.83	25.30	0.748	15.70	5
PM7:Y6	0.90	25.64	0.740	17.04	18
PM6:Y6:J71	0.85	25.55	0.760	16.50	19
PM6:P1:Y6	0.87	25.45	0.732	16.21	20

3.2 Fullerene Derivative Receptor as Third Component

In the past, the fullerene molecule has stood out from many acceptor materials due to its unique three-dimensional spherical structure. However, due to the high symmetry of its wave function, the fullerene and its derivative acceptors have weak light absorption ability, and the absorption band gap is difficult to control. Defects limit the improvement of device efficiency. Some studies have tried to combine fullerene and non-fullerene materials for complementary advantages and improve battery efficiency through fine regulation. Figure 3 is the molecular formula of the fullerene derivative receptor, and Table 3 is the photovoltaic performance parameters of the corresponding device. In 2019, Hou 's research group[21] introduced the fullerene derivative PC61BM

(Figure 3) into the binary mixed system PM6 : Y6.The introduction of PC61BM increased the electron mobility of the device, but also dispersed the Y6 aggregates, reduced the non-radiative energy loss, improved VOC, JSC, and FF, and obtained a device efficiency of 16.50 %. Peng 's group[22] added fullerene derivative PC71BM (Figure 3) as the third component on the basis of PM6 : Y6. The addition of PC71BM promoted the dissociation and generation of charge at the donor / acceptor interface. When mixed in a ratio of 1 : 1.0 : 0.2, the PCE reached 16.67 %. Lu 's group[23] also based on this device, and obtained an efficiency of 16.70 % after mixing in a ratio of 1 : 1.2 : 0.2. Comparing the two groups of experimental results, it can be seen that the performance of the device with the same material composition and different mixing ratios will also be different.

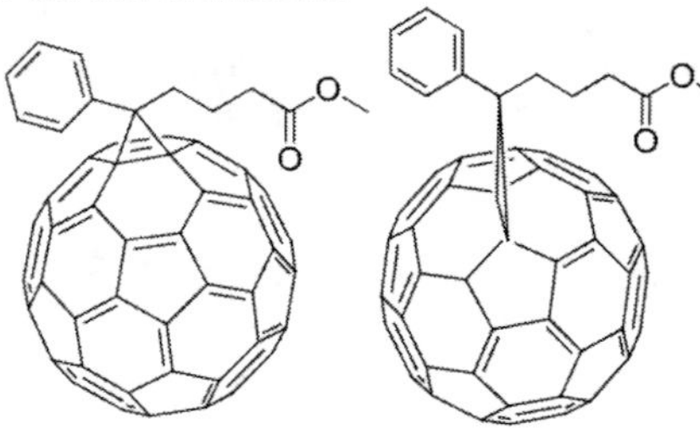

$PC_{61}BM$ **$PC_{71}BM$**

Figure 3. Fullerene derivative receptor molecular formula

Table 3. Photovoltaic performance of ternary organic solar cell devices with fullerene derivative acceptor and Y6 blend

Active layer	V_{OC}(V)	J_{SC}(mA/cm^2)	FF	PCE(%)	ref
PM6:Y6:PC_{61}BM	0.84	25.40	0.770	16.50	21
PM6:Y6:PC_{71}BM	0.85	25.70	0.764	16.67	22
PM6:Y6:PC_{71}BM	0.86	25.10	0.772	16.70	23

3.3 Non-fullerene Small Molecules as the Third Component

The demand for fullerene substitute materials in OSCs has led to the rapid development of non-fullerene receptors. Compared with fullerenes, non-fullerene donors and acceptors have the advantages of better regulation of band gaps and energy levels, simpler synthesis methods, more readily available materials, more easily adjusted structures, and more matching molecules[24]. In order to reduce costs, improve the spectral absorption of organic batteries, improve device stability, and improve battery efficiency[25], the method of introducing non-fullerene molecules as the third component has been proposed and widely used. Figure 4 is the non-fullerene molecular formula, and Table 4 is the photovoltaic performance parameters of the corresponding device. In 2019, Zhan 's group[26] used the ITIC derivative IDIC receptor (Figure 1) as the third component of PM6 : Y6. The structure of IDIC is not only similar to the main body, but also the lowest unoccupied molecular orbital (LUMO) level is higher than that of Y6, thereby increasing the mobility of the device charge and obtaining higher VOC. The PCE also reached 16.51 %. In 2020, Shi 's group[27] designed and synthesized BTP-S2 (Figure 4), and used it to prepare ternary OSCs : PM6 : Y6 : BTP-S2. The device significantly reduces energy loss and promotes charge separation, and the optimal device efficiency reaches 17.43 %. Li and co-workers[28] used highly crystalline molecular donor DRTB-T-C4 (Figure 4) as the third component of PM6 : Y6, DRTB-T-C4 can avoid self-aggregation, enhance and balance charge transport, and optimize electron mobility. The device finally achieved an efficiency of 17.13 %. Lu 's group[29] successfully prepared OSCs based on PM6 : BTR : Y6 by adding liquid crystal small molecule donor BTR (Figure 4). The synergistic effect of BTR and PM6 can finely regulate the morphology of the active layer of the device, making the donor-acceptor binding closer, optimizing the crystallization ability, and finally increasing the PCE of the device to 14.70 %. In 2021, Liu 's group[30] introduced Y6-BO into PM6 : Y6 to maintain the crystalline morphology of the device film through the interaction between the active layer materials, and obtained 17.84 % PCE. In 2022, Zeng and co-workers[31] added ITIC-M (Figure 4) to the PM6 : Y6 blend system, and the prepared three components had a good complementary light absorption effect, and the PCE was as high as 18.13 %.

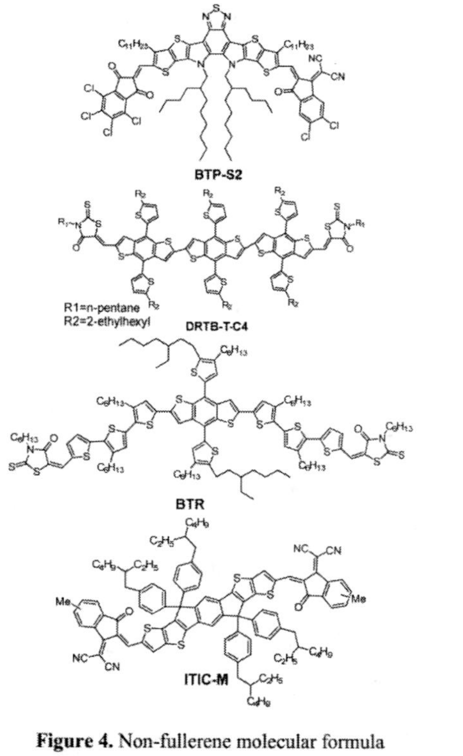

Figure 4. Non-fullerene molecular formula

Table 4. Photovoltaic performance of ternary organic solar cell devices with various non-fullerene molecules blended with Y6

Active layer	V_{OC}(V)	J_{SC}(mA/cm^2)	FF	PCE(%)	ref
PM6:Y6:IDIC	0.87	25.39	0.749	16.51	26
PM6:Y6:BTP-S2	0.88	26.20	0.758	17.43	27
PM6:Y6:DRTB-T-C4	0.85	24.79	0.813	17.13	28
PM6:BTR:Y6	0.83	23.66	0.747	14.70	29
PM6:Y6:Y6-BO	0.84	26.57	0.775	17.30	30
PM6:Y6:ITIC-M	0.86	26.35	0.801	18.13	31

3.4 Solid Additives as the Third Component

The blending morphology of the OSCs active layer plays a key role in improving the PCE of the battery, and adding solid additives to the device is one of the common methods to adjust the morphology and improve the efficiency of the battery. Figure 5 is the molecular formula of solid additives, and Table 5 is the photovoltaic performance parameters of the corresponding devices. In 2020, Li 's group[32] added a volatile solid additive 1,4-diiodotetrafluorobenzene (A3, Figure 5) containing σ-pores to PM6 : Y6-based OSCs. Due to the interaction between halogens, the addition of A3 not only makes the stacking of molecules in the device more orderly and dense, but also promotes the mobility of carriers, which makes the device have better charge transport characteristics, and the PCE reaches 16.50 %. However, studies have shown that even if the concentration of A3 continues to increase to 50mg/mL, the cell efficiency remains at a high level, which also shows that A3 has great application value. Although traditional fluorine-containing solid additives can greatly improve battery efficiency, they cannot be produced on a large scale due to their high toxicity. Therefore, Liu 's group[33] introduced non-toxic halogen-free benzyl benzoate (BB, Figure 5) as an additive into PTB7-Th : PC71BM. It was found that the photovoltaic performance of the device was significantly improved, and the PCE increased from 4.83 % to 9.43 % compared with that without additives.

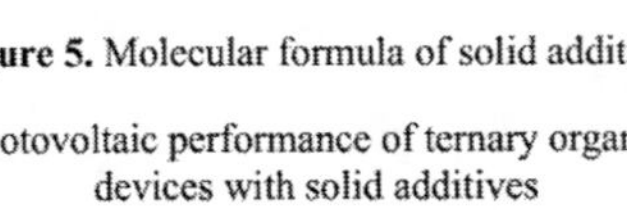

Figure 5. Molecular formula of solid additives

Table 5. Photovoltaic performance of ternary organic solar cell devices with solid additives

Active layer	V_{OC}(V)	J_{SC}(mA/cm^2)	FF	PCE(%)	ref
PM6:Y6:A3	0.86	25.30	0.684	16.50	32
PTB7-Th:PC_{71}BM:BB	0.80	16.74	0.704	9.43	33

3.5 Summary

The emergence of A-DA 'D-A receptor material Y6 has broken the dominant position of ITIC molecules and their derivatives in the field of OSCs. The rigid coplanar alternate push-pull structure of non-fullerene acceptor Y6 enables it to absorb near-infrared light, solve the problem of low near-infrared sensitivity limitation, reduce the band gap and improve the carrier mobility of the device. At the same time, Y6 also has the disadvantages of difficult synthesis, difficult purification and high cost due to the complexity of its central structural skeleton. At present, although the efficiency of Y6-based organic battery devices is close to 19 %, there is no specific requirement for the selection of the third component. In the future, it is still necessary to further optimize the synthesis steps of donor and acceptor materials and explore the working principle of Y6-based batteries.

4. Ternary Blend Organic Solar Cells Based on Polymer Acceptor Perylene Diimide (PDI) And Its Derivatives

PDI (Figure 6) has excellent electron affinity and high electron mobility due to its unique π-π conjugated structure. However, due to the large conjugate plane, the introduction of monomer PDI into the active layer will cause serious molecular self-aggregation, which will inhibit charge diffusion and limit the improvement of battery efficiency. The active sites of PDI can be divided into three types : N, N is called imide, 1,6,7,12 is called Bay, 2,5,8,11 is called α. The paper will introduce groups on the active sites to solve this problem by constructing a non-planar twisted structure. Figure 6 is the molecular formula of PDI and its derivatives, and table 6 is the photovoltaic performance parameters of the corresponding devices.

4.1 Introduction of Simple Groups such as Alkyl Chain or Benzene Ring at Imide, Bay or α Position

Welch 's group[34] synthesized a receptor molecule N-PDI with good film-forming ability by introducing a cetylpyrrole ring at the Bay position (Figure 6). With the addition of this material, the reverse current of the device decreases and the charge mobility increases. The efficiency of the device prepared by blending with PM6 : Y6 reaches 11.20 %. Wasielewskki 's group[35] synthesized Phenyl-PDI receptor (Figure 6) by introducing phenyl on the four α active sites of PDI. After blending with donor PBDTT-FTTE, the device obtained 3.62 % PCE. Zhan 's group[36] used PDI-2DTT (Figure 6) as a receptor material and mixed with PTB7-Th : IDT-2BR. The prepared three components well balance the performance between VOC and JSC, showing stronger energy level transfer, and the best device efficiency can reach 10.10 %.

4.2 PDI was Introduced at Imide, Bay and α Position to Form Polymers

Narayan 's group[37] connected the imide sites of two PDI molecules to synthesize a dimer molecule TP with PDI monomers perpendicular to each other (Figure 6). The device with PBDTTT-C-T as donor and TP as acceptor achieved an efficiency of 3.20 %. Wang 's group[38] used a single bond to connect the inner Bay sites of two PDI monomers, and used Se atoms to connect the Bay sites on both sides of the dimer to form a five-membered ring to obtain the small molecule A1 (Figure 6). The Se atom has a larger electron cloud. After blending with the wide bandgap polymer donor PBDT-T1, the degree of intermolecular orbital overlap is increased, and higher device performance is obtained, with an efficiency of 8.42 %. Li and co-workers[39] modified the two sides of the dimer A1 with cyano groups to obtain the receptor molecule A2 (Figure 6). Although the LUMO energy levels of A2 and PDPP2Tz2T donor molecules are lower than -4.0 eV, the efficiency of the hybrid device still reaches 1.40 %. Studies have shown that the rational construction of PDI dimers is an effective strategy to obtain efficient non-fullerene small molecule receptors.

4.3 Introducing Spacers at Imide, Bay and α Positions

In 2020, Chen 's group [40] synthesized a PDI tetramer $FP_4 3T$ with a fused ring structure by connecting PDI monomers in pairs with thiophene (Figure 6). Due to the influence of steric hindrance effect, there is a large torsion angle between PDI monomers. Therefore, FP43T has strong light absorption and low LUMO energy level. The photovoltaic performance of the device is significantly enhanced after mixing with PTB7-Th, and the cell efficiency reaches 6.05 %. Wang and co-workers[41] synthesized a bb-2PDI dimer (Figure 6) by connecting the Bay sites of two PDI monomers with an oxygen ether bond as a bridge. This structure facilitates more balanced transport of carriers and improves the efficiency of exciton dissociation. The battery efficiency based on PTB7-Th : bb-2PDI is 3.06 %. Li and co-workers [42] synthesized a receptor $PM\text{-}PDI_3$ (Figure 6) containing four PDI molecules by coupling reaction and bromine substitution reaction. Because of its high LUMO level, high exciton dissociation efficiency and good morphology, the PCE reached 7.58 % after blending with PDBT-T1 donor.

4.4 Summary

By modifying the imide, Bay and α positions, researchers aim to improve spectral absorption, adjust HOMO / LUMO energy level matching, and promote electron mobility to optimize the performance of PDI and its derivatives and improve device efficiency. Although the synthesis conditions of PDI derivatives are harsh and the repeatability of the synthesis process is low, they are still concerned by researchers because of their excellent flexibility, high electron mobility and aggregation stability. In the long run, PDI molecules have good application prospects and are expected to make great breakthroughs in the research of non-fullerene small molecule acceptor materials.

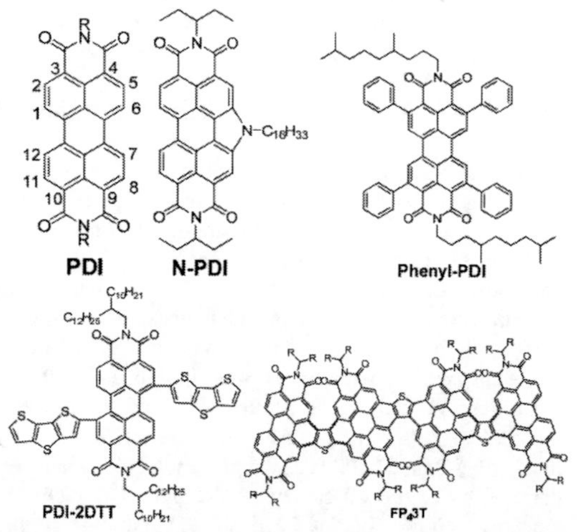

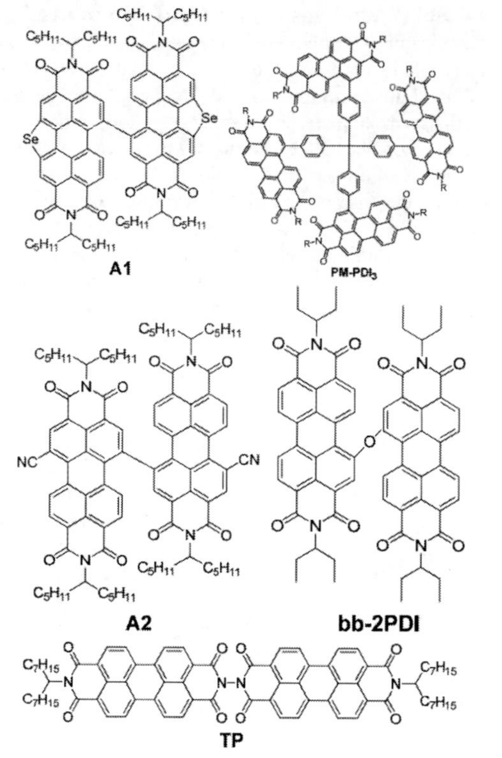

Figure 6. Molecular formula of PDI and its derivatives

Table 6. Photovoltaic performance of ternary blend organic solar cell devices based on PDI and its derivatives

Active layer	V_{OC}(V)	J_{SC}(mA/cm^2)	FF	PCE(%)	ref
PM6:Y6:N-PDI	0.84	23.00	0.580	11.20	34
PBDTT-FTTE:Phenyl-PDI	0.81	9.15	0.484	3.62	35
PTB7-Th:IDT-2BR:PDI-2DTT	1.03	14.50	0.650	10.10	36
PBDTTT-C-T:TP	0.77	9.00	0.460	3.20	37
PBDT-T1:A1	0.95	12.48	0.697	8.42	38
PDPP2Tz2T:A2	0.77	5.50	0.340	1.40	39
PTB7-Th:$FP_4 3T$	0.86	14.70	0.481	6.05	40
PTB7-Th:bb-2PDI	0.81	9.77	0.388	3.06	41
PDBT-T1:$PM\text{-}PDI_3$	0.98	11.02	0.699	7.58	42

5. Summary and Outlook

In this paper, the development status of ternary blend organic solar cells based on ITIC and its derivatives, Y6, perylene diimide (PDI) and its derivatives is introduced in detail. ITIC has good regulation ability. The use of halogenation strategy based on the regulation of the main side chain and end group of ITIC molecules is more conducive to device charge transfer, improving the morphology of the blend film and improving device efficiency. The absorption wavelength range of Y6 molecule is large, which can effectively utilize the light in the near-infrared band and improve the photon capture rate of the device. PDI molecules have many active sites, which can improve the light absorption ability by introducing groups, adjust the LUMO energy level, form a good blend film, and promote the improvement of battery efficiency.

So far, the single-cell efficiency of OSCs has reached 18 %, but there is still a certain gap compared with the commercialized silicon solar cells (25 %). Therefore, the relationship between molecular structure and photovoltaic performance needs to be further explored. How to further optimize the synthesis route is still an important research in this field.

References

1. Zheng Dongzhen, Wang Shaomei. Pollution of petrochemicals on the environment and strategies to cope with it[J]. Contemporary Chemical Research,2022,No.111(10):62-64.
2. Kearns D, Calvin M. Photovoltaic Effect and Photoconductivity in Laminated Organic Systems[J]. The Journal of Chemical Physics, 1958, 29(4):950-951.
3. KROTO H W, HEATH J R, O'BRIEN S C, et al. C60:Buckminsterfullerene[J]. Nature,1985,318(6042):162-163.
4. Lin Y, Wang J, Zhang Z G, et al. An electron acceptor challenging fullerenes for efficient polymer solar cells[J].Advanced materials, 2015, 27(7): 1170-1174.
5. Yuan J, Zhang Y Q, Zhou L Y, et al. Single-Junction Organic Solar Cell with over 15% Efficiency Using Fused-Ring Acceptor with Electron-Deficient Core[J]. Joule, 2019, 3(4):1140-1151.
6. Yang Y, Zhang Z G, Bin H, et al. Side-chain isomerization on an n-type organic semiconductor ITIC acceptor makes 11.77% high efficiency polymer solar cells[J].Journal of the American Chemical Society, 2016, 138(45): 15011-15018.
7. Bin H, Zhang Z G, Gao L, et al. Non-fullerene polymer solar cells based on alkylthio and fluorine substituted 2D-conjugated polymers reach 9.5% efficiency[J].Journal of the American Chemical Society, 2016, 138(13): 4657-4664.
8. Bin H, Gao L, Zhang ZG, Yang Y, Zhang Y, Zhang C, Chen S, Xue L, Yang C, Xiao M, Li Y. 11.4% Efficiency non-fullerene polymer solar cells with trialkylsilyl substituted 2D-conjugated polymer as donor. Nat Commun[J]. 2016 Dec 1;7:13651.
9. Xue L, Yang Y, Xu J, et al. Side chain engineering on medium bandgap copolymers to suppress triplet formation for high-efficiency polymer solar cells[J]. Advanced Materials,2017,29(40),1703344.
10. Fei Z, Eisner F D, Jiao X, et al. An alkylated indacenodithieno [3,2-b] thiophene-based nonfullerene acceptor with high crystallinity exhibiting single junction solar cell efficiencies greater than 13% with low voltage losses[J]. Advanced Materials, 2018, 30(8): 1705209.
11. Xiao Y, Ma R, Zhou G, et al. Ternary blending driven molecular reorientation of non-fullerene acceptor IDIC with backbone order[J]. ACS Applied Energy Materials, 2020, 3(11): 10814-10822.
12. Shi X , Liao X , Ke G , et al. An Electron Acceptor with Broad Visible–NIR Absorption and Unique Solid State Packing for As-Cast High Performance Binary Organic Solar Cells[J]. Advanced Functional Materials, 2018, 28(33),1802324.
13. Zhao W , Li S , Yao H , et al. Molecular Optimization Enables over 13% Efficiency in Organic Solar Cells[J]. Journal of the American Chemical Society, 2017, 139(21):7148.
14. Su D , Pan M A , Liu Z , et al. A Trialkylsilylthienyl Chain-Substituted Small-Molecule Acceptor with Higher LUMO Level and Reduced Band Gap for Over 16% Efficiency Fullerene-Free Ternary Solar Cells[J]. Chemistry of Materials, 2019, 31 (21):8908-8917.
15. Wang H, Liu T, Zhou J, et al. Bromination: an alternative strategy for non-fullerene small molecule acceptors[J]. Advanced Science, 2020, 7(9): 1903784.
16. Cai J, Zhang X, Guo C, et al. Asymmetric and Halogenated Fused-Ring Electron Acceptor for Efficient Organic Solar Cells[J]. Advanced Functional Materials, 2021, 31(31): 2102189.
17. Salleo A, Kline R J, DeLongchamp D M, et al. Microstructural characterization and charge transport in thin films of conjugated polymers[J]. Advanced Materials, 2010, 22(34): 3812-3838.
18. Ruijie Ma,Tao Liu,Zhenghui Luo,et al.Improving open-t voltage by a chlorinated polymer donor endows binary organic solar cells efficiencies over 17%[J].Science China(Chemistry),2020,63(03):325-330.
19. Xie G, Zhang Z, Su Z, et al. 16.5% efficiency ternary organic photovoltaics with two polymer donors by optimizing molecular arrangement and phase separation[J]. Nano Energy, 2020, 69: 104447.
20. Tang Y, Yu J, Sun H, et al. Two compatible polymer donors enabling ternary organic solar cells with a small nonradiative energy loss and broad composition tolerance[J]. Solar RRL, 2020, 4(11): 2000396.
21. Yu R , Yao H , Cui Y , et al. Improved Charge Transport and Reduced Nonradiative Energy Loss Enable Over 16% Efficiency in Ternary Polymer Solar Cells[J]. Advanced Materials, 2019, 31(36):1902302.1-1902302.8.
22. Yan T, Song W, Huang J, et al. 16.67% rigid and 14.06% flexible organic solar cells enabled by ternary heterojunction strategy[J]. Advanced materials, 2019, 31(39): 1902210.
23. Pan M A, Lau T K, Tang Y, et al. 16.7%-efficiency ternary blended organic photovoltaic cells with PCBM as the acceptor additive to increase the open-circuit voltage and phase purity[J]. Journal of Materials Chemistry A, 2019, 7(36): 20713-20722.
24. Zhou Pengchao, Zhang Weidong, Gu Jialu, Chen Huimin, Hu Tengda, Pu Huayan, Lan Weixia, Wei Bin. Efficient organic solar cells based on ternary

non-fullerene system[J]. Journal of Physics,2020,69(19):374-382.

25. Y. Lin, J. Wang, Z.-G. Zhang, et al.An Electron Acceptor Challenging Fullerenes for Efficient Polymer Solar Cells[J]. Advanced Materials, 2015, 27, 1170.
26. Li K , Wu Y , Tang Y , et al. Ternary Blended Fullerene-Free Polymer Solar Cells with 16.5% Efficiency Enabled with a Higher-LUMO-Level Acceptor to Improve Film Morphology[J]. Advanced Energy Materials, 2019, 9(33).
27. Li S, Zhan L, Jin Y, et al. Asymmetric electron acceptors for high-efficiency and low-energy-loss organic photovoltaics[J]. Advanced Materials, 2020, 32(24): 2001160.
28. Li D, Zhu L, Liu X, et al. Enhanced and balanced charge transport boosting ternary solar cells over 17% efficiency[J]. Advanced Materials, 2020, 32(34): 2002344.
29. Tang H, Chen H, Yan C, et al. Delicate morphology control triggers 14.7% efficiency all-small-molecule organic solar cells[J]. Advanced Energy Materials, 2020, 10(27): 2001076.
30. Zhang M, Zhu L, Hao T, et al. Organic Solar Cells: High-Efficiency Organic Photovoltaics using Eutectic Acceptor Fibrils to Achieve Current Amplification[J]. Advanced Materials, 2021, 33(18): 2170142.
31. ZENG, YIHAN, LI, DANQIN, WU, HONGBO, et al. Enhanced Charge Transport and Broad Absorption Enabling Record 18.13% Efficiency of PM6:Y6 Based Ternary Organic Photovoltaics with a High Fill Factor Over 80%[J]. 2022,32(13):2110743.1-2110743.9.
32. Fu J , Chen S , Yang K , et al. A "σ-Hole"-Containing Volatile Solid Additive Enabling 16.5% Efficiency Organic Solar Cells[J]. iScience, 2020, 23(3):100965.
33. Zhang Q , Bao C , Cui S , et al. Boosting the efficiency of PTB7-Th:PC 71 BM polymer solar cells via a low-cost halogen-free supramolecular solid additive[J]. Journal of Materials Chemistry C, 2020,8(46): 16551-16560.
34. Hoff A, Martell M, Gasonoo A, et al. Sidechain-Engineered N-PDIs Processed from Ethyl Acetate as Effective Cathode Interlayers for Organic Solar Cells[J]. Advanced Engineering Materials, 2022: 2201437.
35. Hartnett P E, Margulies E A, Matte H, et al. Effects of Crystalline Perylenediimide Acceptor Morphology on Optoelectronic Properties and Device Performance[J]. Chemistry of Materials, 2016,28(11),3928-3936.
36. Cheng P, Zhang M, Lau T K, et al. Realizing Small Energy Loss of 0.55 eV, High Open-Circuit Voltage >1 V and High Efficiency >10% in Fullerene-Free Polymer Solar Cells via Energy Driver[J]. Advanced Materials, 2017, 29(11),1605216.
37. Shivanna R , Shoaee S , Dimitrov S , et al. Charge generation and transport in efficient organic bulk heterojunction solar cells with a perylene acceptor[J]. Energy & Environmental Science, 2014, 7(1):435-441.
38. D Meng, D Sun, C Zhong,et al. High-Performance Solution-Processed Non-Fullerene Organic Solar Cells Based on Selenophene-Containing Perylene Bisimide Acceptor[J].Journal of the American Chemical Society, 2016, 138(1): 375-380.
39. Yu Y, Yang F, Ji Y, et al. A perylene bisimide derivative with a LUMO level of−4.56 eV for non-fullerene solar cells[J]. Journal of Materials Chemistry C, 2016, 4(19): 4134-4137.
40. Chen H, Wang L, Sun H, et al. PDI-based heteroacenes as acceptors for fullerene-free solar cells: importance of their twisted geometry[J]. New Journal of Chemistry, 2020, 44(30): 13093-13099.
41. Wang H, Chen L, Xiao Y. Oxygen-ether-bridged perylene diimide dimers: Efficient synthesis, properties, and photovoltaic performance[J]. Dyes and Pigments, 2020, 180:108508.
42. Li G , Yang W , Wang S , et al. Methane-perylene diimide-based small molecule acceptors for high efficiency non-fullerene organic solar cells[J]. Journal of Materials Chemistry C, 2019, 7.

https://doi.org/10.1051/matecconf/202338001015

Numerical Analysis on Natural Vibration of Cylindrical shell with Different Cross-section

Peng Wang [1], Qingyong Niu [2], Min Liu [1], Zhao Li [1], Xiaoming Cao [1], Hao Zhang [1,*]

[1] Wuhan Second Ship Design and Research Institute, Wuhan 430205, China
[2] College of Science, Wuhan University of Science and Technology, Wuhan, 430065, China

Abstract. Cylindrical shell in engineering has different cross-section, i.e., circular section, oval section, rectangular section, and so on. In this work, numerical analysis is conducted on natural vibration of cylindrical shell with different cross-section, and the difference is studied between modal frequency and modal shape from different section shell. Numerical results show that, with the same length, width /thickness and boundary condition, the first 10 order modal frequency from circular section shell is nearly 60% bigger than that from the rectangular section shell. This indicates that the load carry ability of circular cylindrical shell is much better than rectangular cylindrical shell, which will help to support the design of cylindrical shell in engineering.

1. Introduction

Cylindrical shell has wildly applications in engineering, such pipeline, water tank and pressure hull of submarine. But the section of cylindrical shell may be different

For the vibration of circular cylindrical shell, much work has been done. Arnold and Warburton [1] applied the energy method to establish the governing equation of natural vibrato circular cylindrical shell. Leissa [2] summarized recent works on free vibration of cylindrical shell, and he also gave the suitable range for different methods. It's a great work. Zhang [3] uses the wave propagation approach to analyze the free vibration of circular cylindrical shell with finite shell length.

For vibration of non-circular cylindrical shell ,available research work is a bit fewer. Klosner and Pohle [4-5] studied the natural vibration of oval cylindrical shell with both ends simply-supported. Sewall [6] conducted series experiments to study the vibration of oval cylindrical shell, and experimental data was plotted in figures and tables for researchers. Baudak [7] applied spline matching method to investigate the free vibration of non-circular cylindrical shell, and modal frequency was calculated. Suzuki and Leissa [8-9] constructed a new Lagrange function, and use it to study the vibration of non-circular cylindrical shell with thickness varying along circumferential direction.

In this work, numerical analysis is conducted on natural vibration of cylindrical shell with different cross-section, and the difference is studied between modal frequency and modal shape from different section shell.

2. Numerical Simulation

2.1 Research object

There are two research objects in this work, one is circular cylindrical shell, the other is rectangular cylindrical shell. The shell length is 8000mm, thick ness is 10mm. Radius of circular section is 2600mm, the length and width of rectangular section are also 2600mm with fillets of 800mm.

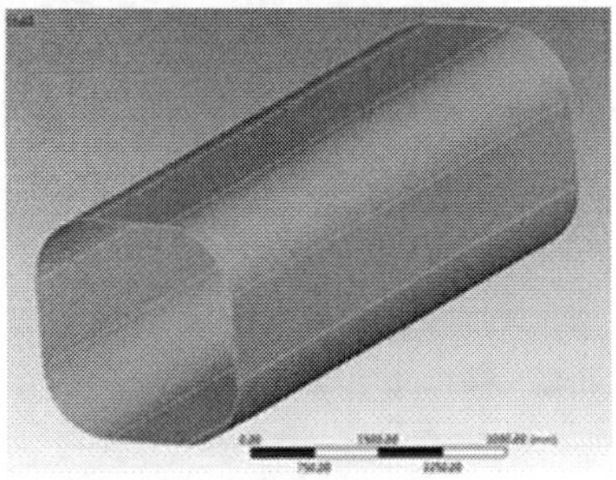

(a) rectangular

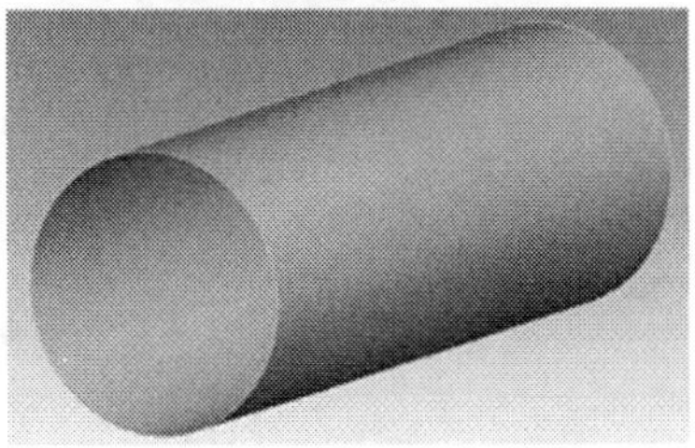

(b)circular

Fig. 1 Shell model

* Corresponding author: zhyf1408@126.com

2.2 Boundary condition

In numerical simulation, the boundary condition of both circular and rectangular cylindrical shell is clamped for both shell ends.

2.3 Materials

In numerical simulation, material of shell model is 022Cr17Ni12Mo2 steel, parameters of material is shown in Table.1. The density is 7850 kg/m^3, Elastic module is 200GPa, the Poisson's ratio is 0.3, the Yield strength is 177MPa.

2.4 Model Mesh

The mesh work for shell model is conducted in ANSYS software, the element size is 50mm.

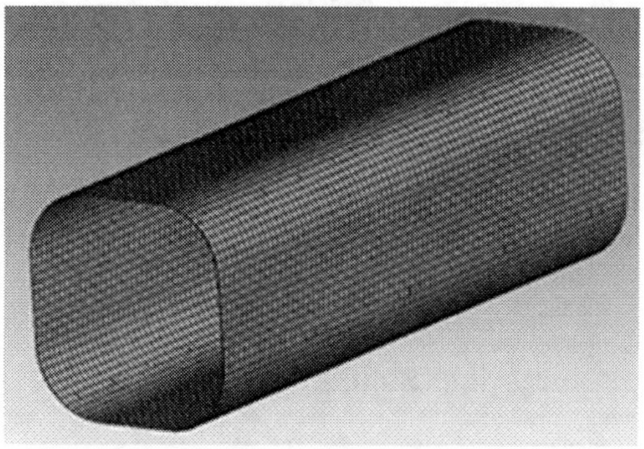

(a) rectangular

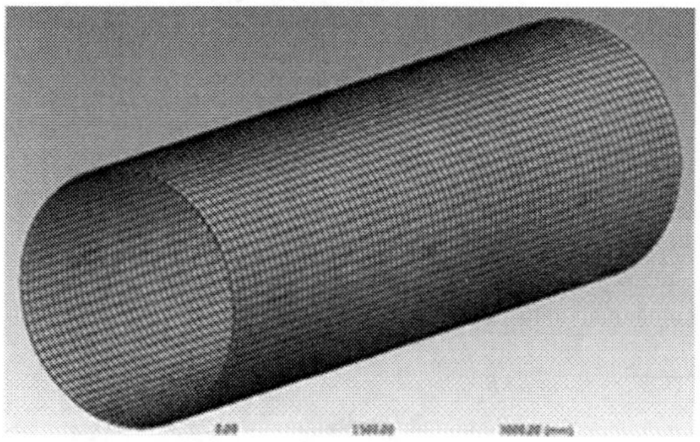

(b) circular

Fig. 2 Mesh model

3. Comparative analysis of modal parameters

Comparative analysis is conducted for modal parameters of cylindrical shell with different sections, results are thown in Table 2. The difference of modal frequency is defines as

$$dif = \frac{f_{cir} - f_{rec}}{f_{rec}} \times 100\%$$

Where, f_{cir} and f_{rec} are respectively modal frequency from circular and rectangular cylindrical shell.

It's noted that, compared to rectangular cylindrical shell, model frequencies from circular cylindrical shell is much bigger. The increase is more than 60%, and it varies with change of mode order. Furthermore, increase of the seventh mode order is the biggest for the first ten modes, and it is nearly 127.6%. This phenomenon could be explained, , and circular section has good load carry ability in circumferential direction, while the rectangular section has curvature mutation and flexural stiffness will decrease, and modal frequency will finally be reduced. Therefore, circular section could be applied to replace rectangular section cylindrical shell to enhance structural load carrying ability or increase structural model frequency.

Table 1. Comparative analysis on modal frequency

Mode order	f_{rec}	f_{cir}	*dif*
1	15.66	29.46	88.1%
2	17.53	29.46	68.0%
3	17.53	34.78	98.4%
4	20.28	34.78	71.5%
5	20.62	37.17	80.3%
6	22.06	37.17	68.5%
7	22.06	50.20	127.6%
8	24.27	50.20	106.9%
9	25.95	51.89	99.9%
10	26.25	51.89	97.7%

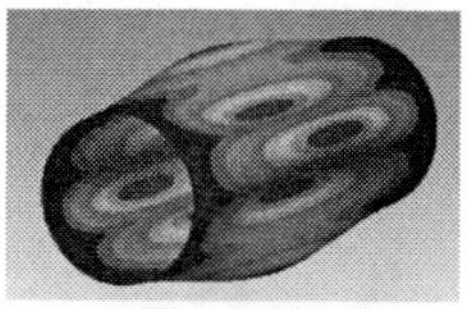

First mode First mode

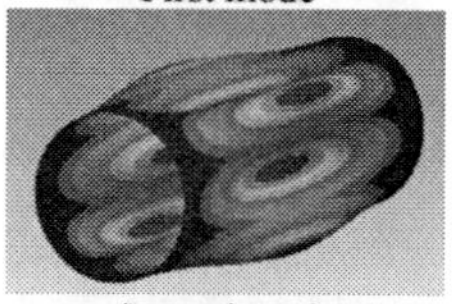

Second mode Second mode

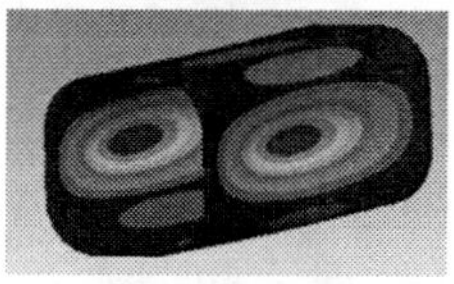
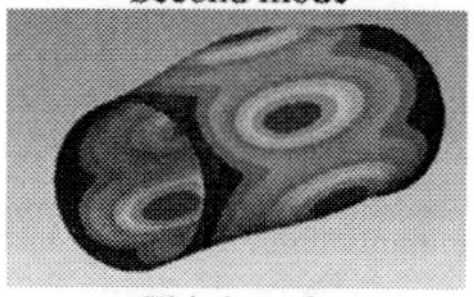

Third mode Third mode
(a) rectangular section (b) circular section

Fig. 3 the first four mode shapes of circular and rectangular cylindrical shell

4. Conclusion

In this work, numerical analysis is conducted on natural vibration of cylindrical shell with different cross-section, and the difference is studied between modal frequency and modal shape from different section shell.
Numerical results show that, with the same length, width /thickness and boundary condition, the first 10 order

modal frequency from circular section shell is nearly 60% bigger than that from the rectangular section shell. This indicates that the load carry ability of circular cylindrical shell is much better than rectangular cylindrical shell. Besides, modal shape from circular and rectangular shell in the same order is also obvious different.

Conclusions in this work will help to support the design of cylindrical shell in engineering.

Acknowledgements

We sincerely thank the financial support from the Hubei Provincial Natural Science Foundation (No. ZRMS2020002231) and the Young Top-notch Talent Cultivation Program of Hubei Province.

References

1. Arnold R N, Warburton G B, Flexural vibration of the walls of thin cylindrical shells having freely supported ends. Proceedings of the Royal Society. 1949, Serial A,197: 238-256.
2. Arnold R N, Warburton G B. The flexural vibration of thin cylinders. Proc. Inst. Mech. Engineers. 1953, Serial A, 167: 62-80
3. Leissa A W. Vibration of Shells. Washington; NASA,1973,SP-288.
4. Zhang X M, Liu G R, Lam K Y. Vibration analysis of thin cylindrical shells using wave propagation approach. Journal of Sound and Vibration. 2001, 239(3): 397-403.
5. Klosner J M, Pohle F V. Natural Frequencies of an Infinitely Long Noncircular Cylindrical Shell. Polytechnic Institute of Brooklyn, N.Y., 1958,PIBAL Rept. 476.
6. Klosner J M. Frequencies of an Infinitely Long Noncircular Cylindrical Shell.Polytechnic Institute of Brooklyn, N.Y., 1959, PIBAL Rept.552.
7. Sewall J L, William M T, Christine G P. An experimental and analytical vibration study of elliptical cylindrical shells. NASA, 1971, TN D-6089.
8. Budak V D, Grigorenko A Y, Puzyrov S V. Free vibrations of the non-circular cylindrical shells. The 10th International Conference "Modern Building Materials, Structures and Techniques". 2010, 874-879.
9. Suzuki K, Leissa A W. Free vibration of noncircular cylindrical shells having circumferential varying thickness. Journal of Applied Mechanics. 1985, 52: 149-154.
10. Suzuli K, Leissa A W. Exact solutions for the free vibrations of open cylindrical shells with circumferentially varying curvature and thickness. Journal of Sound and Vibration. 1986, 107(1): 1-15.

 https://doi.org/10.1051/matecconf/202338001016

Defect Detection of Aluminum Profiles based on Improved Feature Pyramids

Jie Wang, YanSha Zhang, Feng Pan, Lin Wang *

College of Data Science and Information Engineering Guizhou Minzu University, Key Laboratory of Pattern Recognition and Intelligent System of Guizhou Province, Guiyang, China

Abstract. For the surface defects of aluminum profiles, there are problems of multi-scale, small object and irregular shape. This paper proposes a defects detection algorithm based on improved feature pyramid. This method compresses and saves the feature information extracted by the backbone networks, and calculates the similarity between deep and shallow features, so as to alleviate the phenomenon of loss of feature information and weakening of feature expression ability, thereby solving the problem of multi-scale and small object. At the same time, deformable convolution is introduced to enhance the feature extraction ability of the model and alleviate the detection problems caused by irregularly shaped defects. To verify the effectiveness of the proposed method, Faster R-CNN was used as the basic detection algorithm to conduct ablation experiments, and compared with the classical detection algorithm, the accuracy rate was as high as 72.8%. The experimental results show that the proposed method has a good performance on the task of aluminum profile defects detection, and is superior to the comparative detection algorithms.

Keywords: aluminum profile defects detection; feature pyramid networks; faster r-cnn; deformable convolution networks.

1. Introduction

Aluminum profiles are not only widely used in infrastructure construction, but they are also integral to aerospace, automobile manufacturing, and other industries. However, various defects will appear on the surface of the aluminum profiles during production and transportation. Therefore, it is crucial to improve quality by detecting surface defects in the aluminum profiles. Traditional surface defects detection methods are typically completed by human eyes and have the disadvantages of slow speed and low accuracy, failing to accomplish the task of the actual production process while increasing production costs and affecting the economic benefits of enterprises.

In recent years, Convolutional neural networks (CNN) has performed well in the field of computer vision, which has attracted the attention to scholars in various fields. CNN automatically learns image features of low-level to high-level at different levels, which solves the defects of the time-consuming and labor-intensive manual design of feature extraction operators in traditional techniques. Compared with traditional computer vision technology, the deep learning method does not require tedious preprocessing process, and the extracted features are more generalized due to the model's self-learning. Therefore, some object detection algorithms with excellent performance based on CNN have been proposed. such as R-CNN1, SSD2, YOLO3, which promote the application of CNN in surface defects detection. Liao4 improves YOLOv4 for surface defects detection on printed circuit boards (PCBs), Zhang5 used the attention mechanism to effectively integrate with the SSD detector to complete the surface defects detection of rare earth magnetic materials, Wang6 and Xiang7 simultaneously uses K-means clustering and feature pyramid networks to optimize Faster R-CNN, thereby achieving surface defects detection of aluminum profiles.

In summary, although many scholars have conducted in-depth research and application on surface defects detection, there are still two problems:

(1) In the face of multi-scale and small object problems with surface defects, most researchers' solutions are to use Faster R-CNN8 and feature pyramid networks (FPN)9 combination to solve, but the FPN has design flaws, which reduces the representation ability of feature and loses feature information.

(2) The shape of the surface defects is irregular. Feature extraction networks increase the receptive field by stacking convolutional layers, but the receptive field is usually a regular grid that cannot match the target shape, so background information is mixed in during feature extraction.

In view of the above problems, the following two aspects of work have been done:

* Corresponding author: wanglin@gzmu.edu.cn

(1) An information enhancement feature pyramid networks (IE-FPN) is designed to alleviate the information loss problem and the weakening of multi-scale feature characterization ability in the FPN by constructing information re-injection and information similarity modules.
(2) By introducing a deformable convolutional networks (DCN)10, the receptive field of the feature extraction networks can conform to the shape of surface defects, reducing the focus on background information. As well as the area and proportion distribution of the annotation boxes in the statistical training set to generate appropriate anchors.

2. Related work

In the task of defects detection, it is particularly important to accurately detect surface defects. Due to the excellent performance of the two-stage detection algorithm, the Faster R-CNN is selected as the basic framework in this paper. Faster R-CNN is composed of feature extraction networks (backbone networks), region of interest pooling (ROI Pooling) and prediction head. The structure of Faster R-CNN is shown in Figure 1.

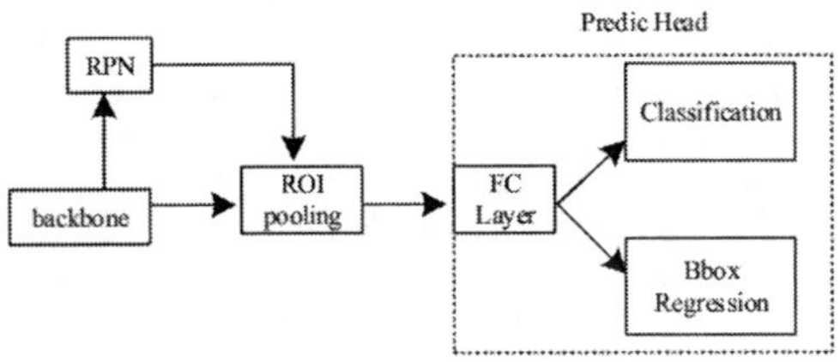

Figure 1. The structure of Faster R-CNN

Compared with natural images, surface defects images have small objects and multi-scale problems. However, the Faster R-CNN only uses the last layer feature map output by the backbone networks of prediction, which cannot effectively solve the problem. This is because when the backbone network's structure is deepened, the down-sampling rate will also increase, which means that the resolution of the feature maps will be reduced, resulting in the loss of the detailed information about the detection targets on the low-level features, and the detection effect of small areas of defects is not effective.
To alleviate the above two problems and improve the surface defects detection effect. We introduce two new modules in Faster R-CNN, including: (1) Improved feature pyramid networks, which maps defects of different scales to corresponding feature maps for prediction, which has alleviated the detection performance poor caused by small targets and multi-scale problems. (2) DCN, by applying a coordinate offset to the ordinary convolution kernel, the convolution kernel can better adapt to the shape of the defects. Considering the model complexity, DCN is only introduced into the last two stages of backbone networks.
In the feature extraction process, low-level features have rich location and detailed information, which is conducive to the detection of small-area targets, while high-level feature maps have rich semantic information and are suitable for handling large-size targets. Therefore, FPN maps objects of different scales to feature maps of different receptive fields, and then uses the corresponding feature maps for prediction. The definition of FPN can be expressed in equation (1).

$$P_i = \begin{cases} W_{3\times3}(W_{1\times1}(C_i) + Unsample(P_{i+1})) & i=1,2,3 \\ W_{1\times1}(C_i) & i=4 \end{cases} \quad (1)$$

Where $W_{3\times3}$ and $W_{1\times1}$ are 3×3 convolution and 1×1 convolution respectively, and Unsample is bilinear interpolation. The structure of FPN is shown in Figure 2.

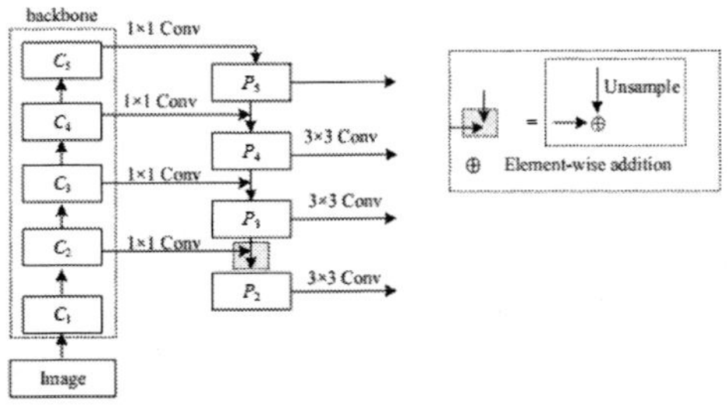

Figure 2. The structure of FPN

3. Defects Detection of Aluminum Profiles based on Improved Feature Pyramids

3.1 Information Enhancement Feature Pyramid Networks

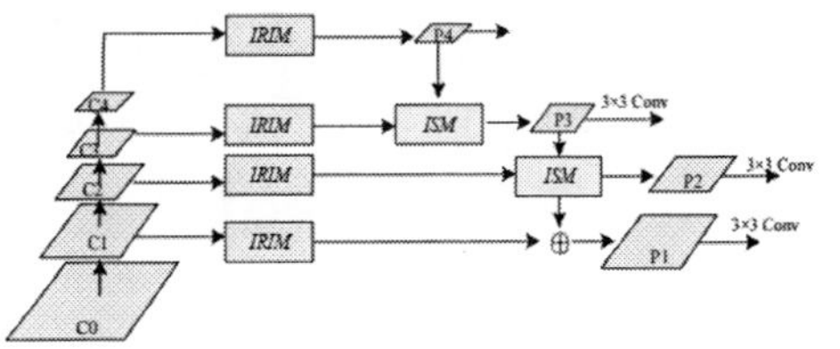

Figure 3. The structure of IE-FPN

The feature pyramid networks has been widely used in surface defects detection due to its structural and performance advantages. However, AugFPN11 points out that there are two design flaws: (1) There is a semantic gap between the feature semantic information about different levels, and direct fusion will weaken the characterization ability of multi-scale features; (2) The lateral connection in the FPN uses 1×1 convolution to reduce the channel dimension of the feature, which will lead to information loss. To alleviate information loss and the weakening of the representation ability of multi-scale features, we propose IE-FPN. The structure of IE-FPN is shown in Figure 3.
IE-FPN introduces the information re-injection module (IRIM) and the information similarity module (ISM) based on FPN. The feature information extracted from the backbone networks is passed through the IRIM module to

alleviate the problem of information loss caused by channel dimension reduction. Considering that the similarity matrix between the calculated features is affected by the spatial size of the features, this paper uses the features of P4 and P3, P3 and P2. The ISM module is introduced into fusion to alleviate the problem of reduced representational ability brought about by the semantic gap between feature fusion. The IE-FPN can be defined as follows.

$$P_i = \begin{cases} W_{3\times3}(IRIM(C_i) + Unsample(P_{i+1})) & i=1 \\ W_{3\times3}(IRIM(C_i) + ISM(P_{i+1})) & i=2,3 \\ IRIM(C_i) & i=4 \end{cases} \quad (2)$$

Where i is the level index of the feature map, and $W_{3\times3}$ represents the convolution operation of size 3×3. The following is a detailed introduction to the two modules of IRIM and ISM.

The information re-injection module aims at the information loss caused by dimension reduction of horizontal connection channels in FPN. By compressing the channel information of each pixel on the unreduced feature map into space, the compressed information feature map of the unreduced feature map is obtained, and then the compressed information feature map is normalized to achieve the functions of information filtering and information preservation. Feature map before channel dimension reduction. The framework of IJIM is shown in Figure 4.

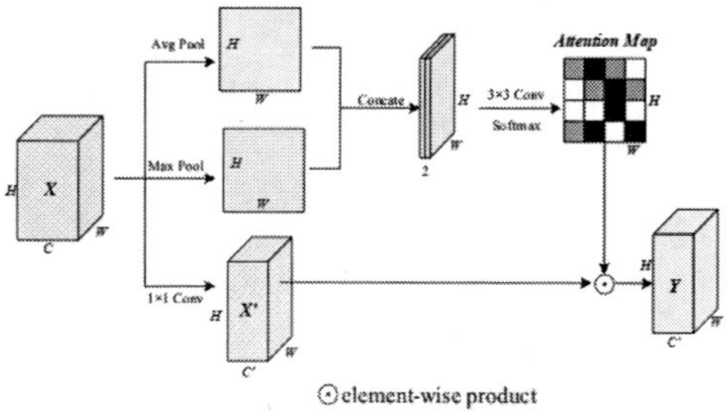

Figure 4. The framework of IJIM

The input feature map $X \in \mathbb{R}^{H\times W\times C}$ is the original feature output by the backbone networks, and the feature map X will be operated through three branches. Firstly, the pooling operation is performed according to the channel dimension to obtain the information compression feature map with the dimension $H\times W\times 1$. Considering that single pooling may lead to information loss, the maximum and average pooling parallel methods are adopted to alleviate the loss. Secondly, the two information compression feature maps are spliced for convolution to achieve the effect of information interaction, and then normalized by SoftMax to obtain the spatial attention map. At the same time, the feature map X is dimensionally reduced by 1×1 convolution to obtain the feature map $X' \in \mathbb{R}^{H\times W\times C'}$, and finally the dimension-reduced feature map X' is multiplied by the corresponding element of the spatial attention map to obtain the output feature Y. The IRIM can be defined as follows.

$$Y = W_{X'}.X \Box \sigma(W_P[AvgPool(X);MaxPool(X)]) \quad (3)$$

$$AvgPool(X_{i,j}) = \frac{\sum_{k=1}^{C} X_{i,j}^{k}}{C} \quad (4)$$

$$Maxpool(X_{i,j}) = \max X_{i,j}^{k} \quad (5)$$

Where $W_{X'}$ is the 1×1 convolution parameter, W_P is the 3×3 convolution parameter, σ is the normalization function SoftMax, C is the number of channels of the feature map X, and k is the index value.

In summary, we propose IRIM compresses the channel information into space by combining two pooling operations to obtain the information compression feature map, which is normalized and then applied to the feature map after channel dimension reduction, which alleviates the original problem. The problem of information loss after feature channel dimension reduction. At the same time, in the spatial domain, the response of the feature map with important information is strengthened, and the quality of the feature map is improved.

The design of the information similarity module is mainly aimed at the problem that the semantic information asymmetry exists in the feature fusion of different levels in FPN, and the direct fusion will lead to the weakening of the feature expression ability. ISM reduces the information gap between feature maps by calculating the pixel similarity between features at different levels, and reduces the information gap between feature maps through the similarity matrix, thereby alleviating the weakening of feature representation. The framework of ISM is shown in Figure 5.

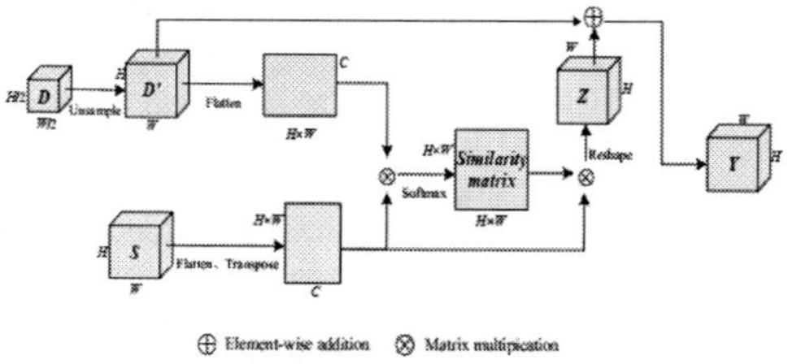

Figure 5. The framework of ISM

In the figure, $D \in \mathbb{R}^{(H/2)\times(W/2)\times C}$, $S \in \mathbb{R}^{H\times W\times C}$ represent the deep features and shallow features extracted by the feature extraction networks, respectively. Up-sampling is performed so that the spatial size of the feature map is consistent with the shallow layer to obtain the feature map $D' \in \mathbb{R}^{H\times W\times C}$. Flatten the feature map D' and S, perform matrix multiplication and normalize SoftMax to obtain the similarity matrix, apply the similarity matrix to the shallow features to obtain the feature map $Z \in \mathbb{R}^{H\times W\times C}$. Finally, the feature map Z and the corresponding pixels of D' are added and fused as the output feature Y. The ISM can be defined as follows.

$$Y^k_{i,j} = \sum_{t=1}^{H\times W} (S^k_t M_{t,i}) + D'^k_{i,j} \tag{6}$$

Where i and j are represented as the horizontal and ordinate coordinates index of the feature map, and k is the index value of the number of channels C. M is a similarity matrix, and its expression is as follows.

$$M_{t,i} = \frac{\exp(\sum_{k=1}^{C} D'^k_t S^k_i)}{\sum_{i=1}^{H\times W} \exp(\sum_{k=1}^{C} D'^k_t S^k_i)} \tag{7}$$

In summary, inspired by the attention mechanism, IE-FPN constructs IRIM and ISM to alleviate the problem of information loss in feature pyramids and the weakening of feature representation capabilities.

3.2 Deformable Convolution Networks

Currently, most detection models use ordinary convolutions for feature extraction. It uses the regular slide window R to sample on the input feature map X, and then the parameter values on the sliding window of the sampling point domain are weighted to obtain the output feature map Y, and the ordinary convolution is calculated as shown in the equation (8).

$$Y(p_0) = \sum_{p_n \in \mathbb{R}} W(p_n) \cdot X(p_0 + p_n) \tag{8}$$

Where p_0 represents the pixel position on the output feature map Y, W represents the parameter value on the sliding window, p_n represents the coordinates of all parameter values on the sliding window R, taking the sliding window of 3×3 as an example, p_n= {(-1, -1), (-1,0),...,(0,1), (1,1)}.

However, there is a problem of inconsistent shape of surface defects, and if ordinary convolution is used for stacking, the receptive field of the model will not adapt to the shape of the target, resulting in the networks model failing to extract valid information from the target or extracting background interference information. Therefore, we use deformable convolution to alleviate the above problems. Deformable convolution calculates the coordinate offset of the sliding window at the sampling point on the feature map by using the convolution operation branch. This makes the sampling point no longer a regular square, but changes with the shape of the target. Deformable convolution is calculated as shown in the equation (9).

$$Y(p_0) = \sum_{p_n \in \mathbb{R}} W(p_n) \cdot X(p_0 + p_n + \Delta p_n) \tag{9}$$

Where Δp_n is the predicted coordinate offset of the sample point. Figure 6 shows a schematic diagram of ordinary convolution and deformable convolution of 3×3 size.

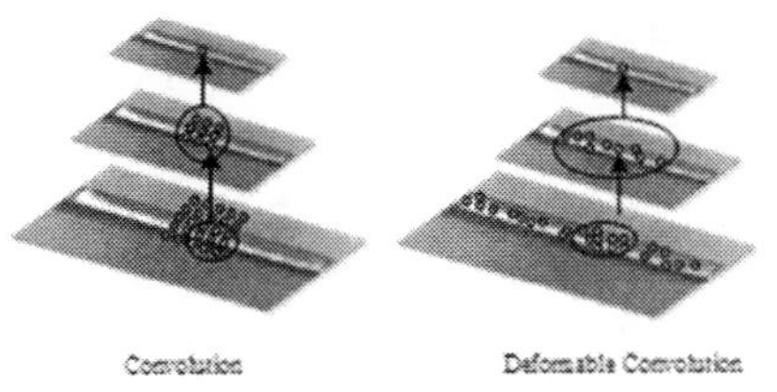

Figure 6. The Schematic diagram of ordinary convolution and deformable convolution

The red box in the Figure 5 is the defect area. For the target with a large aspect ratio, the ordinary convolution will extract the background information in the process of extracting features. By improving the grid sampling method of ordinary convolution, spreading the sampling points can make the feature extraction networks more suitable for the shape of the target. Therefore, deformable convolutions can capture more challenging and irregularly shaped defects.

4. Experiments

4.1 Datasets and Settings

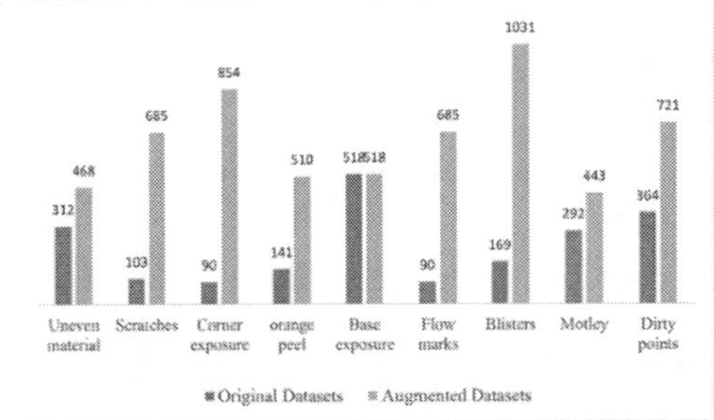

Figure 7. The sample distribution of the training set

This paper uses the datasets of surface defects of aluminum profiles with annotations from the Alibaba Cloud platform to conduct experiments. The datasets have a total of 2092 single-defect samples with a resolution of 2560 × 1920, and the defects have 9 aluminum profile defect categories. The datasets are divided into training set and validation set according to 8:2. However, there are problems such as excessive and uneven categories in the training set, which cause troubles in the model training process. To solve the problem, we adopt offline enhancement of the training set, and combine operations such as rotation and image sharpening to achieve the purpose of data enhancement. Figure 7 shows the sample distribution before and after data augmentation, and it can be seen that the above problems are alleviated.

The experiment is based on the Pytorch framework. The GPU is NVIDIA GeForce RTX 3060, the memory is 12GB.The CPU is Intel(R) Core(TM) i5-11400f.

4.2 Implementation Details

The optimizer used by the networks model when updating the parameters is Adam, and the weight decay factor in the optimizer is 0.0001. Set the initial Learning Rate to 0. 0001, batch size is 4, and using a fixed step learning rate attenuation strategy, the learning rate attenuation rate (Lr scheduler) is 0.1, and a total of 30 iterations (epoch) are trained. At the same time, because the image resolution of the datasets is too large, it requires a large amount of computation, so the input image size will be scaled by 1/4 of the original resolution when training and validating the networks. The performance of the aluminum profile defects detection model is evaluated by using the mean average precision (mAP).

4.3 Ablation Experiments

To better set the model parameters, improve the performance of the algorithm. In this paper, a statistical analysis of the aspect ratio of the training set annotation boxes and the proportion of defective areas is performed. Figure 8 shows the aspect ratio distribution of the annotation boxes, and Table 1 shows the proportion of defect area.

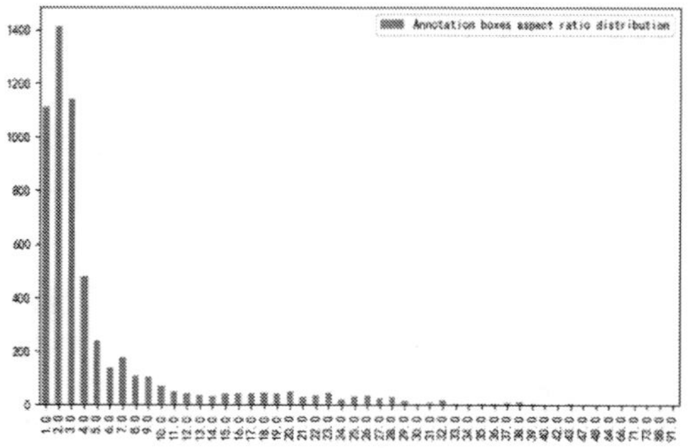

Figure 8. the aspect ratio distribution of the annotation boxes

Table 1. Training set defect area distribution

Defect area ratio	Sample proportion
<0.01%	0.03%
0.01%~0.1%	7.73%
0.1%~1%	14.2%
1%~10%	38.87%
10%<	39.15%

According to the definition of the International Society of Optical Engineering, if the size of the object is less than 0.12% of the original drawing, it can be considered a small object. Table 1 can conclude that there are 7.76% small objects in the training set. For the Faster R-CNN algorithm, the aspect ratio parameter for generating the default box is [0.5,1,2] and the area parameter is [(8*feature map down-sampling rate)^2], it is difficult to adapt to the aspect ratio and area of the dimension box, thus limiting model performance. Therefore, this paper adjusts the default generation box parameters, and at the same time considers the avoidance of increasing the amount of computation, so that the corresponding feature map generates 10 anchor per pixel, the aspect ratio parameter is [0.14, 0.5, 1, 2, 7], and the area parameter is [(4*feature map down-sampling rate)^2, (8*feature map down-sampling rate)^2].

To verify the effectiveness of the proposed IE-FPN algorithm, this paper introduces each module to the basic detector in turn for ablation experiments. The steps are as follows: (1) the backbone networks is Resnet5012 and the Faster R-CNN of the FPN is used as the basic detector; (2) the IRIM is introduced into the FPN in the basic detector; (3) the IRIM and the ISM are introduced at the same time on the basis of the basic detector, that is, the original FPN is replaced by the IE-FPN proposed in this paper; (4) on the basis of step 3, the aspect ratio and area parameters selected after the analysis of the annotation boxes are introduced; (5) and the DCN is finally introduced to obtain the final detector algorithm of this paper. The mAP results of each experiment are shown in Table 2.

Table 2. Performance Comparison of Different Algorithms

Experiment number	IRIM	ISM	Appropriate Anchor ratio	DCN	mAP (%)
(1)	×	×	×	×	59.3
(2)	✓	×	×	×	64
(3)	✓	✓	×	×	64.6
(4)	✓	✓	✓	×	70.9
(5)	✓	✓	✓	✓	**72.8**

As can be seen from the Table 2, the mAP detected by the basic detection algorithm is 59.3%. Replacing the FPN with the IE-FPN proposed in this paper, that is, the mAP of experiment 3 is increased by 5.3% compared with the basic detection algorithm experiment. The mAP of the final model obtained using all the improved strategies reached 72.8%, an increase of 13.5% over the basic detection algorithm. In summary, the introduction of other optimization strategies such as IRIM and ISM can improve the performance of basic detection algorithms and improve the detection effect of surface defects of aluminum profiles.

To further verify the effectiveness of the method proposed in this paper, it is compared with the representative object detection algorithm based on deep learning at this stage. The comparison results are shown in Table 3.

As can be obtained from Table 3, the proposed method has the best detection performance compared with the current popular object detection algorithm. Among them, compared with the SSD512 algorithm, mAP of the detection obtained by our method is increased by 12.5%; the average detection accuracy improved by 7.8%, 6.3%, and 6.8% compared to the Cascade R-CNN13+FPN algorithm, the Retinanet14+FPN algorithm, and the YOLOv315 algorithm. Meanwhile, comparing the detection speed of different algorithms by fps, our method processes 18.2 images per second, which is faster than Cascade R-CNN+FPN algorithm and slower than Retinanet+FPN, YOLOv3 and SSD512. This is a normal result because the basic detection framework in this paper is Faster R-CNN, which improves detection accuracy at the expense of detection speed.

Table 3. Performance Comparisonr of Mainstream Algorithms

Method	mAP(%)	FPS
Cascade R-CNN+FPN	65	17.7
Retinanet+FPN	66.5	25
YOLOv3	66	32.8
SSD512	60.3	29.8
Ours	**72.8**	**18.2**

5. Conclusion

Aiming at the surface defects of aluminum profiles, there are problems of multi-scale, small objection and irregular shape of defects, this paper proposes a surface defects detection method based on the improved FPN. By constructing and introducing IRIM and ISM, the information loss and semantic gap of FPN are alleviated. Secondly, the DCN is introduced to make the receptive field of the feature extraction networks conform to the shape of the surface defects of aluminum profile, reducing the attention of background information. Experiments on the surface defects datasets of aluminum profiles verify the effectiveness of the method, and the detection accuracy of the proposed method is better than that of the current popular algorithm of defects detection. In order to make the defects on the surface of aluminum profiles more in line with industrial scenarios and improve the quality of products, we will continue to study the surface defects detection of aluminum profiles in the future, in the hope of better identification accuracy and detection speed.

References

1. Girshick R, Donahue J, Darrell T, et al, "Rich feature hierarchies for accurate object detection and semantic segmentation," Proceedings of the IEEE conference on computer vision and pattern recognition. 2014, pp. 580-587.
2. Liu W, Anguelov D, Erhan D, et al, "Ssd: Single shot multibox detector," European conference on computer vision. Springer, Cham, 2016, pp.21-37.
3. Redmon J, Divvala S, Girshick R, et al, "You only look once: Unified, real-time object detection," Proceedings of the IEEE conference on computer vision and pattern recognition. 2016, pp. 779-788.
4. Liao X, Lv S, Li D, et al, "YOLOv4-MN3 for PCB Surface Defect Detection," Applied Sciences, 2021, 11(24): 11701.
5. Zhang B, Fang S, Li Z, "Research on Surface Defect Detection of Rare-Earth Magnetic Materials Based on Improved SSD," Complexity, 2021, doi: 10.1155/2021/4795396.
6. Wang J, Meng Z H. "Deformable Feature Pyramid Network for Aluminum Profile Surface Defect Detection," Journal of Physics: Conference Series. IOP Publishing, 2020, 1544(1): 012074.
7. K. Xiang, S. Li, M. Luan, Y. Yang, "Aluminum product surface defect detection method based on improved Faster RCNN," Chin. J. Sci. Instrum, 2021, pp. 191-198.
8. S. Ren, K. He, R. Girshick, J. Sun, "Faster r-cnn: Towards real-time object detection with region proposal networks," Advances in neural information processing systems, 2015, 28, doi:10.48550/arXiv. 1506.01497.
9. T. Y. Lin, P. Dollár, R. Girshick, K. He, B. Hariharan, S. Belongie, "Feature pyramid networks for object detection," Proceedings of the IEEE conference on computer vision and pattern recognition. 2017, pp. 2117-2125.
10. J. Dai, H. Qi, Y. Xiong, et al, "Deformable convolutional networks," Proceedings of the IEEE international conference on computer vision. 2017, pp. 764-773.
11. C. Guo, B. Fan, Q. Zhang, S. Xiang, C. Pan, "Augfpn: Improving multi-scale feature learning for object detection," Proceedings of the IEEE/CVF conference on computer vision and pattern recognition. 2020, pp. 12595-12604.
12. K. He, X. Zhang, S. Ren, J. Sun, "Deep residual learning for image recognition," Proceedings of the IEEE conference on computer vision and pattern recognition. 2016, pp. 770-778.
13. Z. Cai, V. Nuno, "Cascade r-cnn: Delving into high quality object detection," Proceedings of the IEEE conference on computer vision and pattern recognition. 2018, pp. 6154-6162.
14. T. Y. Lin, P. Goyal, R. Girshick, K. He, D. Piotr, "Focal loss for dense object detection," Proceedings of the IEEE international conference on computer vision. 2017, pp. 2980-2988.
15. J Redmon , A Farhadi, "Yolov3: An incremental improvement," arXiv preprint arXiv:1804.02767, 2018, doi: 10.48550/arXiv. 1804.02767.

https://doi.org/10.1051/matecconf/202338001017

Study of Unconfined Compressive Experiment on the Geogrid Flexible Reinforced Compound

Jintian Tong [1], Wei Wang [2]

[1]Wuhan University of Engineering Science, Wuhan, China
[2]Beiqiao Zhongqi Design Co., Ltd. Wuhan Branch, Wuhan, China

Abstract. In recent years, reinforced technique has been widely used in civil engineering technology. In order to deeply understand the behavior of the reinforced structure, this paper use the geogrid as the reinforced materials. The unconfined compressive tests were operated in the laboratory. The specimens were made by sharp sand which was classified according to the density at ρ=1.55g/cm3, ρ=1.75g/cm3, ρ=1.85g/cm3.Each specimen was reinforced with the layers of 0,1,2,3,4,5. Through analysis the experimental data, it's founded that the density reinforced materials and the layers have an effect on the unconfined compressive strength. The specimen with bigger density and more reinforced layers has a better compressive strength. Different reinforced materials have different reinforced effect.

Key words: geogrid; reinforced; unconfined compressive.

Nowadays, the use of reinforcement technology is becoming more and more extensive. In order to obtain a large number of systematic mechanical property test data of flexible reinforced soil complex, systematically sort out and analyze the test results, and extract the basic mechanical properties of flexible reinforced soil complex, this paper conducts an unconfined compressive test of flexible composite reinforced soil. The experiment mainly studies the influence of geotextile, the number of layers of reinforcement, and the density of soil on the compressive strength of soil. Provide some guidance for later practical projects.

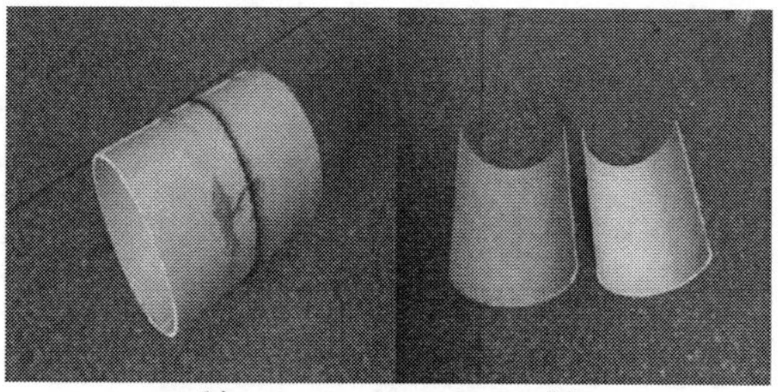

(a) (b)

Figure 1. (a) plastic, (b) two halves plastic

Figure 2. latex film

1. Preparation for the unconfined compression test

1.1 Test instruments

There are test instruments required for unconfined pressure test: A Lu Qiangyi. The inner diameter is 152mm.One plastic round tube with a height of 189mm.And split into two halves, as shown in Figure 1(a)(b). The diameter is 150mm.There are several latex films with a height of 240mm, as shown in Figure 2.A pad with a diameter of 150mm and a thickness of 39mm.Several rubber bands, bench scales, shovels, stopwatches.

1.2 Test materials

The test sand used in this paper is pure sand, and the sand grains with particle size greater than 1.18mm in the sand are removed. The particle size distribution is shown in Figure 4 below. The geogrid adopts a two-way geogrid.

1.3 Test the reinforcement scheme

The test adopts indoor test to study the influence of the density of the sample, the reinforced material and the number of reinforced layers on the unconfined compressive strength. The height of the specimen is 150mm, and the number of layers reinforced is n=0, 1, 2, 3, 4, 5.

1.4 Preparation of test specimens

Before the test, the yellow sand for the test is baked in the oven for several hours, the moisture is dried, and then it is screened with a screening machine to remove the particle size greater than 1.18mm, and the particle size of the sand is measured. Tests of one density specimen at a time are performed. In accordance with the reinforcement scheme, the reinforcement is sequential. When preparing a specimen of this density, the mass of sand required for each density and the mass of sand required for each height of each layer under each reinforcement scheme are first calculated, and the specimen forming is the same except that the mass of the sand weighed in each layer is different. First of all, the latex film should be put on the bottom pad and hooped with a rubber band, and the plastic pipe should be hooped up with a rubber band on the cushion block, and the excess part of the latex film should be turned over on the plastic tube. There are rubber band hoops up to ensure fixation.

Weigh the mass of each layer of sand and load it into a latex film with a plastic tube on the outside. Then the geotextile that has been cut in advance is laid with reinforced material. However, when the reinforcement is 0 layer, it is also best to add sand in layers. In this way, each layer is as uniform as possible. Start loading to keep the sand loose before tamping it to the height of each layer. After the test is formed, the outer plastic sleeve is removed, and then the unconfined compression test is carried out. The initial specimen is shown in Figure 3.

Figure 3. initial specimen figure

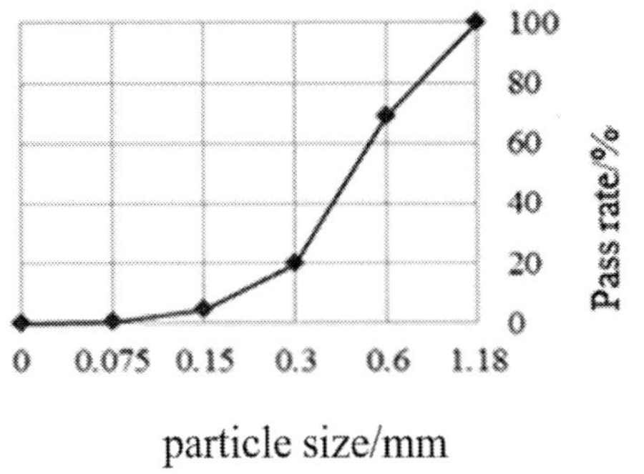

Figure 4. particle size distribution figure

2. Unconfined compression test of geogrids

According to the data recorded in the test and the calculation results, a stress-strain (q-ε) relationship diagram was made. Q represents the compressive stress at the vertical pressure to which the specimen is subjected, ε represents the compressive strain over time. Take about ε=15% for the strain of sample failure. According to the relationship diagram, the change law of the strength of the flexible reinforcement complex at different densities and different number of reinforcement layers is analyzed. The geogrid is shown in Figure5, and the curve is shown in Figure 6 to Figure8.

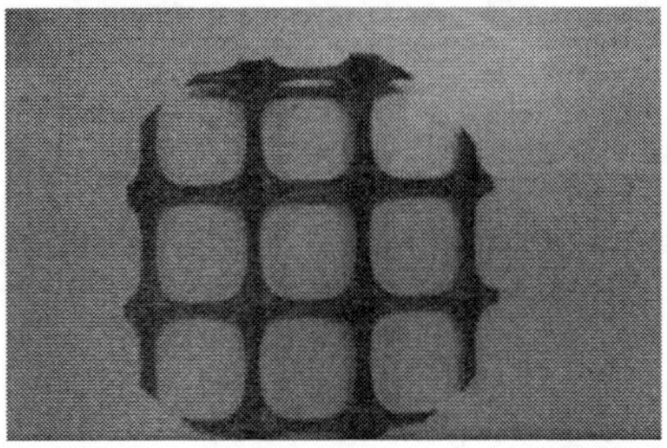

Figure 5. geogrid

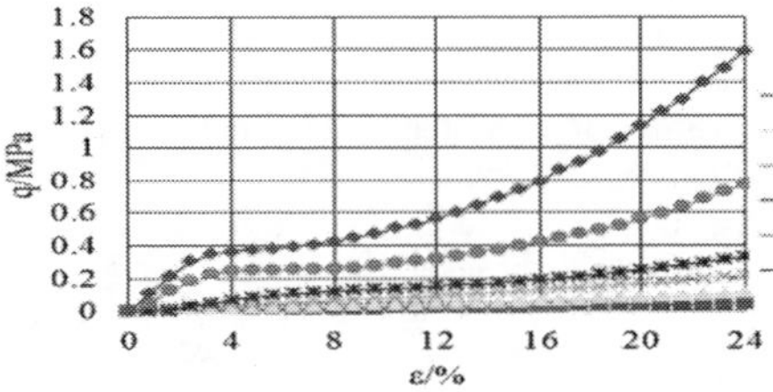

Figure 6. ρ=1.55g/cm^3 q-ε graph

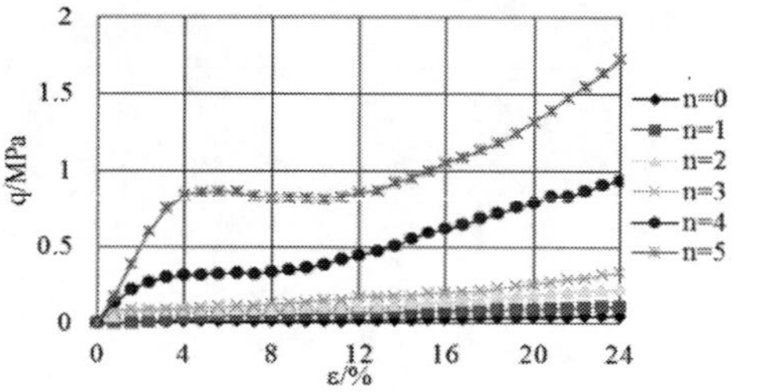

Figure 7. ρ=1.75g/cm^3 q-ε graph

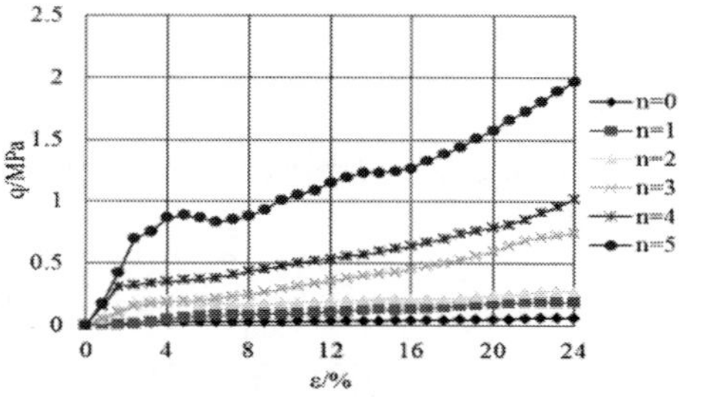

Figure 8. ρ=1.85g/cm^3 q-ε graph

According to Figures 6 to 8, we can get:
(1) As the number of layers of the geogrid increases, the strength of the specimen increases layer by layer. In particular, the more layers are added, the greater the pressure value. Especially when the number of layers stiffened is n = 5, the pressure value of the three densities of the specimen at the time of failure strain increases by about 30 times compared with the pure sand. This shows that the number of layers reinforced by the geogrid has a certain influence on the strength of the specimen, which can enhance the unconfined compressive strength of the specimen, and the compressive strength of the unreinforced specimen is improved.
(2) At n=1, 2, the reinforcement effect of the geogrid is not so obvious, and the pressure does not increase very much in the process of increasing the strain, and the pressure value at the time of failure is about 4 times the pressure of the unadded geogrid. But it is still less than 0.5KN. However, when the number of layers of the geogrid is n=3, 4, 5, the pressure value increases significantly. When ρ=1.55g/cm^3 the sample is n=3, 4, 5 layers, the pressure value at the time of failure is 8, 17, 33 times without geogrid. When ρ=1.75g/cm^3 specimen is in n=3, 4, 5 layers, the pressure value at the time of failure is 6, 19, 33 times without geogrid. When ρ=1.85g/cm^3 the specimen is in n=3, 4, 5 layers, the pressure value at the time of failure is 10, 15, 30 times without geogrid. It shows that excessive density does not promote the increase of specimen strength. Therefore, it is necessary to choose the appropriate density according to the actual engineering situation.
(3) In the above 3 graphs, the curve of strain between ε=4%-8% is almost horizontal, and the pressure value does not change. I think it may be because each layer of the specimen is not very evenly pounded. The top layer is denser. So the pressure value of the sample increases with the increase of strain. But the lower layers may not be as dense as the top layer, and the sample is looser. So there is a period in the process of increasing strain that is that the pressure of the sample does not change much at ε=4%-8%.

3. Conclusion

Based on all the previous analyses, summarizing the number of layers of reinforcement, the difference of reinforcement materials and the influence of the density of the sample on the strength of the flexible complex, the author believes that the essence of the above phenomenon is the cause of the reinforcement mechanism. Now the more popular reinforcement mechanism is friction reinforcement mechanism and quasi-cohesion theory [1-3]. Because the geogrid is a geotextile with holes, the friction with the sand is reduced. If the density is very small and the sand body is very loose, such as ρ=1.55g/cm^3, then the integrity of geogrid and sand formation is certainly not as good as geotextile and sand body formation. The higher the density of the specimen, the denser it is, and the reinforced material can form a whole, stability and integrity are good. Geogrid reinforcement is a good reinforcement material for flexible composite that improves the compressive strength of sand, and is a good flexible support structure.

References

1. LIU Yuzhuo. Highway engineering Ruan Ji treatment [M]. People's Communications Press, 2003.
2. ZHOU Zhigang. Analysis on the application of reinforced soil technology in geotechnical engineering [J]. Shanxi Architecture, 2010, 9.
3. GAO Xiang. Application research on the design method of reinforced soil structure[D].Master's thesis, Southeast University, 2001, 5.

Design of double-layered framed plate with equivalent impedance

Zhao Li[1], Hao Zhang[1], Qingyong Niu[2], Peng Wang[1,*], Xiaoming Cao[1]

[1] Wuhan Second Ship Design and Research Institute, Wuhan 430205, China
[2] College of Science, Wuhan University of Science and Technology, Wuhan, 430065, China

Abstract. To get clear the vibration mechanism of sandwith shell using the experiment method, a double-layered plate is designed for the sandwich shell by matching the equivalent impedance. Numerical analysis is conducted, and In 10Hz-4000Hz, the differenceof impedance from double-layered plate and sandwich shell should be less than 10%. A new design is given for the double-layered plate, and which is quite simple for construction. Then, the double-layered plate could be applied to conduct series vibration tests to get clear the influence of sandwich core material on structural vibration, while it could also reduce the costs of money and construction time.

1. Introduction

Sandwich structures have wild application in engineering, double-layered plate or shell is one kind of sandwich structures, it consists of two plates, the up and down plate and the sandwich cores is frame.

Many works have been done about the mechanic, vibration and dynamic performance of sandwich structures.

Chen et.al.[1] Analyzed the free vibration of sandwich structures with different lattice type. Lu et.al.[2] Studied the influence of random damage on dynamic performance sandwich plate with metal core, with the application of finite element model. Wu et.al.[3] Conducted system work for mechanic performance of very light sandwich plate with frictional pyramid core.

Besides, sandwich structure could be applied to reduce structural vibration and noise level by selecting reasonable core material [4-5]. For a sandwich shell, the final design is to be made for good vibration and noise reduction. The best way to select suitable sandwich core for the shell is to construct a series of shell model filled with different sandwich cores. But this will cost much money and time, due to the difficulty of construction for sandwich shell. Therefore, a double-layered plate is designed with equivalent impedance to the sandwich shell, so as to lay foundation for future work on sandwich cores selection. Since the time and money cost for double-layered plate is relatively much smaller.

In this work, to study the influence of sandwich core on vibration of sandwich cylindrical shell, a double-layered plate design is put forward by using the impedance equivalent method. In the first step, three different double-layered plate designs are given for the sandwich cylindrical shell, and the structural impedance are calculated and compared with that of the sandwich cylindrical shell. With comparative analysis, a good double-layered plate design is selected, which will lay foundation for future work to study the influence of sandwich core on vibration of sandwich cylindrical shell.

2. Design process of double-layered plate

2.1 Target model

The sandwich cylindrical shell is the target model here, and the geometry is shown in Fig.1. The shell is of length 690mm, thickness of the outside plate is 3.2mm, the frame is 2mm, and the inner plate is 2.7mm.

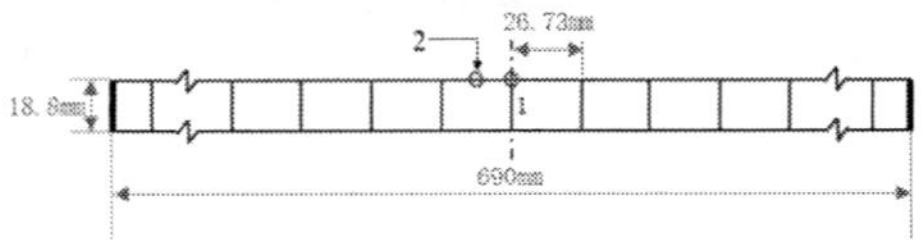

Fig. 1 half section of double-layered cylindrical shell

During the numerical simulation, there are two excitation points shown in Fig.1 as points 1 and 2.

2.2 Material

The material of sandwich cylindrical shell is titanium alloy, elastic module is 108GPa, density is 4510kg/m^3, the Poisson's ratio is 0.34.

The material of double-layered plate is steel, elastic module is 210GPa, density is 7850kg/m^3, the Poisson's ratio is 0.3.

2.3 Criterion

During the numerical simulation, the difference of impedance from double-layered plate and sandwich

* Corresponding author: 1025500148@qq.com

cylindrical shell should be less than 10% in frequency range, 10Hz-4000Hz.

2.4 Multi-plans for double-layered plate

1) Plate 1
Structural parameters o plan 1 is the same to those of sandwich cylindrical shell, so thick of the up plate is 3.2mm, the down plate is 2.7mm, and the frame and plates at the ends are 2mm.
2) Plate 2
Some structural parameters in plan 2 has changed for easy construction, thickness for the up and down plate is 3mm, the frame and plates at the ends are 2mm. The other parameters are the same with the sandwich shell.
3) Plate 3
Some structural parameters in plan 3 is different, thickness for the up plate is 2mm, the down plate is 3mm, the frame and plates at the ends are 2mm. The other parameters are the same with the sandwich shell.

2.5 Equivalent impedance analysis process

1) Construct the numerical model of sandwich shell and double-layered plate with the CATIA software, according to the parameters in plans.
2) Numerical analysis pre-process, such as meshing, boundary condition setting, using the Hypermesh software.
3) Harmonic response analysis, using the LMS virtual.Lab software.
4) Numerical results process, obtain the acceleration response and impedance of shell and plates.
5) Comparative analysis on impedance from different plans to the sandwich shell, discuss and get some conclusions.

3. numerical results discussion

Comparative analysis on impedance curves from plates and sandwich shell against excitation frequency at point 1 is plotted in Fig.2, and impedance curves at point 2 is plotted in Fig.3, and the difference in 10Hz-4000Hz is shown in table X.

It's noted from Fig.2 that at point 1: impedance from different plates varies up and down around that from the sandwich shell; fluctuation from plate 1 is minimum, plate 3 is the second, and plate 2 is the maximum.

At point 2: impedance from different plates also varies up and down around that from the sandwich shell; fluctuation from plate 2 is minimum, plate 1 is the second, and plate 3 is the maximum.

For the impedance difference: At point 1, plate 1 is minimum, plate 3 is the second, and plate 2 is the maximum, and they are all less than 10%; At point 2, plate 2 is minimum, plate 1 is the second, they are both less than 10%, and plate 3 is the maximum, 28.5%, which is much bigger than 10%.

In a word, plate 2 is selected as the equivalent plate for the sandwich shell.

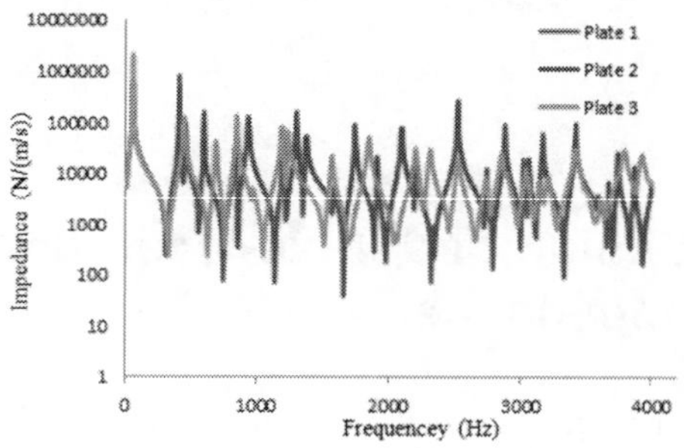

Fig. 2 impedance curves from different plates at point 1

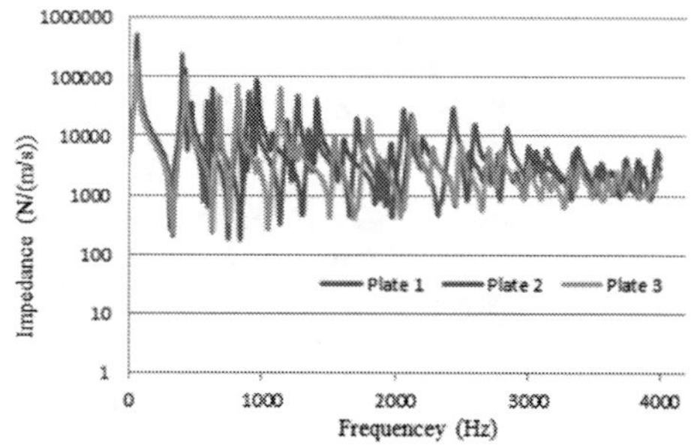

Fig. 3 impedance curves from different plates at point 2

Table 1. Difference of impedance from different plates compared with that from sandwich shell (10Hz-4000Hz)

Difference	Plate 1	Plate 2	Plate 3
Point 1	7.21%	9.85%	-7.75%
Point 2	4.51%	2.84%	-28.56%

4. Conclusions

In this work, a double-layered plate is designed for the sandwich shell using the equivalent impedance method. In 10Hz-4000Hz, the impedance of double-layered plate and sandwich shell is nearly the same with that of sandwich shell, and the difference is less than 10%. Then, the double-layered plate could be applied to conduct series vibration tests to get clear the influence of sandwich core material on structural vibration, while it could also reduce the costs of money and construction time. This work will lay foundation for the vibration tests of sandwich shell with different cores in the future.

Acknowledgements

We sincerely thank the financial support from the Hubei Provincial Natural Science Foundation (No. ZRMS2020002231) and the Young Top-notch Talent Cultivation Program of Hubei Province.

References

1. Chen J E, Zhang W, Sun M, et al. Free vibration analysis of composite sandwich plates with different truss cores[J]. Mechanics of Advanced Materials and Structures, 2018, 25(9): 701-713.
2. Lu L L, Song H W, Huang C G. Effects of random damages on dynamic behavior of metallic sandwich panel with truss core[J]. Composites Part B: Engineering, 2017, 116:278-290.
3. Wu Q Q, Ashkan V, Mohamad E A, et al. Lattice materials with pyramidal hierarchy: Systematic analysis and three dimensional failure mechanism maps[J]. Journal of the Mechanics and Physics of Solids, 2019, 125:112-144.
4. Gui-Lan Yu and Hong-Wei Miao. On vibration isolation of sandwich plate with periodic hollow tube core[J]. Journal of Sandwich Structures and Materials, 2019, 21(3): 1119-1132.
5. Qiang Chenxu et al. Bandgaps and vibration isolation of local resonance sandwich-like plate with simply supported overhanging beam [J]. Applied Mathematics and Mechanics, 2021, 42(11): 1555-1570.

Influence of Frame Weld on Mechanic Performance of Double-walled Shell

Peng Wang[1], Zhao Li[1], Qingyong Niu[2], Hao Zhang[1,*], Xiaoming Cao[1]

[1] Wuhan Second Ship Design and Research Institute, Wuhan 430205, China
[2] College of Science, Wuhan University of Science and Technology, Wuhan, 430065, China

Abstract. Weld type between parts will make a difference on structural performance, which is especially important for a daouble-walled shell. To get clear this problem, numerical simulation is conducted to investigate the influence of weld plans between frame and out shell on structural stress and stability. Numerical results show that the change of weld plan between frame and out shell is small in case of a 15MPa external pressure, stress of shell is below the critical value, and the ultimate bearing capacity changes a little. Thus, the weld plan of point weld for frame and out shell could be applied while the double-walled shell is under construction.

1. introduction

Double-walled shell is one typical structure of ship structure, and it has widely application in ship design, submarine structures.

Many works have been done on the characteristics of double-walled shell. Zhang et.al. [1] analyzed the effect of perforation on the sound transmission through a double-walled cylindrical shell. Then, Zhang et.al. [2] extended their work to study the Sound transmission through micro-perforated double-walled cylindrical shells lined with porous material. Xie et.al. [3] give an analytical method to predict the vibration of double-walled cylindrical shell. Lee and Kim [4] conducted an experiment to get clear the Sound transmission through a double-walled cylindrical shell. Cacciolati [5] established a theoretical model for direct sound transmission loss of double-walled shell. Recent works are mainly on the vibration and sound radiation of Double-walled shell.

Besides, in the view of construction, the complex structure elements make its construction to be quite difficult. In ideal case, the outside shell, frames and inner shell should be welded together perfectly. But, in fact, the weld work for Double-walled shell is very hard. Usually, frames are firstly welded to inner shell. Then the weld between frame and outside shell is to be done, which is especially difficult. To solve this problem, point weld along the frame could be applied, if the change makes little difference on mechanism of Double-walled shell.

In this work, to investigate the influence of frame weld on mechanic of double-walled cylindrical shell, numerical simulation is conducted on in case of different frame weld plans, comparative analysis on shell stress and stability is also carried out. This work will lay foundation for the building and construction of Double-walled shell.

2. Numerical model

2.1 geometry

The Double-walled cylindrical shell is shown in Fig.1. the shell is of length 720mm, it's made up of the out shell, frame and inner shell. The out shell is 3.2mm thick, the inner shell is 2.7mm. The frame is 2mm thick, the height is 18.8mm, and distance is 26.67mm, and 25 circular frames are concerned in the model. One half sphere of 15mm thickness is attached to each ends of the shell.

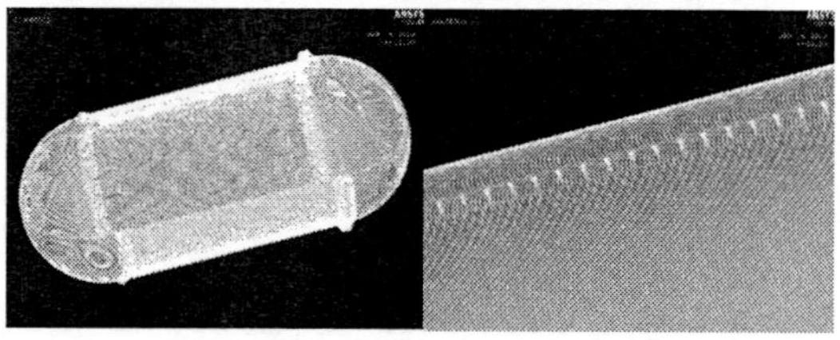

Fig. 1 Model for the double-walled shell

2.2 frame weld plans

1) all weld

In this case, the frame is all welded to out shell. During the numerical simulation, frame and out shell is set to be bounded, and node merging is applied in other words.

2) point weld

In this case, the frame is point welded to out shell, and diameter of wed point is 3mm, and 8 weld point is distributed along the circular direction at each frame. In numerical simulation, at the weld point, the out shell and frame is node merging, and no separation is applied a other place besides the weld points, and the friction factor is set to be 0.25.

* Corresponding author: zhyf1408@126.com

3) contact only
In this case, the frame is not welded to the out shell, and friction contact only is applied. During numerical simulation with ANSYS, contact between frame and out shell is friction contact, and the friction factor is 0.25.
Material for Double-walled shell is titanium alloy, the elastic module is108Ga, the Poisson's ratio is 0.3, the yield stress is 865MPa.

3. Numerical results

3.1 All weld

In the case of all weld, the stress of out shell, frame and inner shell is displayed in Fig.2 under a external hydrostatic pressure 15MPa, the stability curves is also given in Fig.3.
It's noted that the maximum stress in out shell, frame and inner shell is 781MPa , structure is safe under 15MPa external hydrostatic pressure.
It's also noted from Fig.3 that the ultimate bearing capacity of Double-walled shell in case of all weld is 19.96MPa.

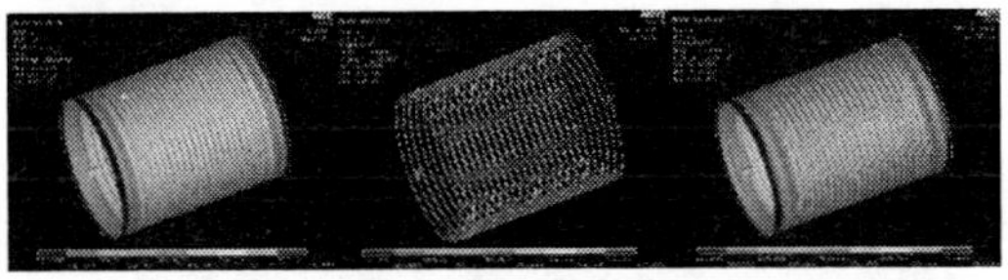

Fig. 2 stress of Double-walled shell under 15MPa external pressure in case of all weld (a)out shell; (b) frame; (c)inner shell

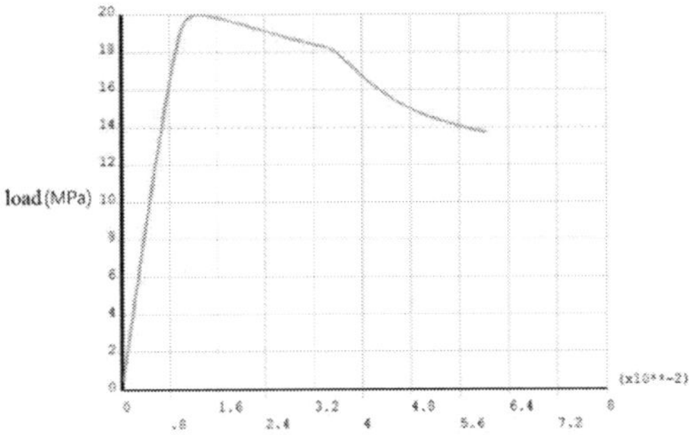

Fig. 3 load versus deformation curves in case of "all weld"

3.2 Point weld

In the case of point weld, the stress of out shell, frame and inner shell is displayed in Fig.4 under 15MPa external hydrostatic pressure, the stability curves is also given in Fig.5.
It's noted that the maximum stress in out shell, frame and inner shell is 840MPa, structure is safe under 15MPa external hydrostatic pressure.
The ultimate bearing capacity of Double-walled shell in case of point weld is 19.64MPa.

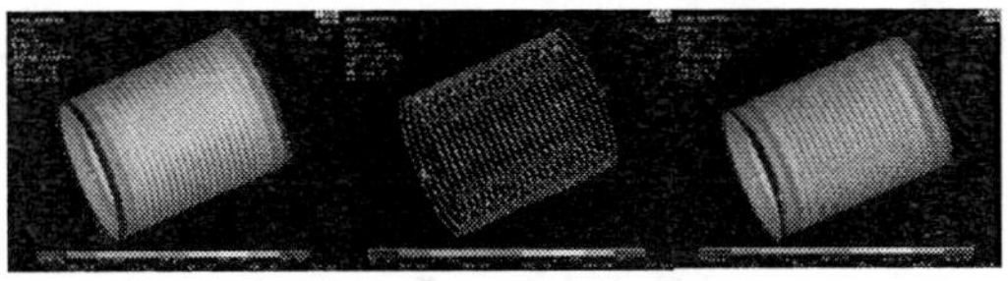

Fig. 4 stress of Double-walled shell under 15MPa external pressure in case of point weld (a)out shell; (b) frame; (c)inner shell

3.3 Contact only

In the case of contact only, the stress of out shell, frame and inner shell is displayed in Fig.5 under 15MPa external hydrostatic pressure.
It's noted that the maximum stress in out shell, frame and inner shell is 783MPa, structure is safe under 15MPa external hydrostatic pressure.
The ultimate bearing capacity of Double-walled shell in case of contact only is 19.60MPa.

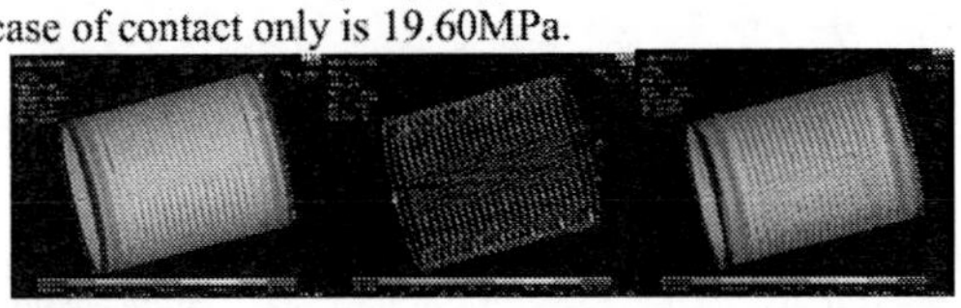

Fig. 5 stress of Double-walled shell under 15MPa external pressure in case of contact only (a)out shell; (b) frame; (c)inner shell

3.4 Comparative analysis

Focused on the stability of double -layered shell, the ultimate bearing capacity of shell under external pressure is given in Table 1.
It's noted that, compared with the stability n the case of all weld, the change of weld plan on stability of shell is small, the ultimate bearing capacity of shell in case of point weld is 19.64MPa, the reduction is 1.59%; the ultimate bearing capacity of shell in case of point weld is 19.60MPa, the reduction is 1.77%.
Therefore, contact only or point weld plan could be applied while the Double-walled shell is under construction.

Table 1. Comparative analysis on stability of shell in different weld plans

parameter	All weld		Point weld		Contact only	
	P_{cr}	Reduction	P_{cr}	Reduction	P_{cr}	Reduction
Stability	19.96	-	19.64	1.59%	19.60	1.77%

4. Conclusions

In this work, numerical simulation is conducted to investigate the influence of weld plans between frame and out shell on structural stress and stability. It's shown that the change of weld plan between frame and out shell is small, under 15MPa external pressure, shell is safe in all cases, and the ultimate bearing capacity changes a little. Thus, the weld plan of point weld or contact only for frame and out shell could be applied while the Double-walled shell is under construction.

Acknowledgements

We sincerely thank the financial support from the Hubei Provincial Natural Science Foundation (No. ZRMS2020002231) and the Young Top-notch Talent Cultivation Program of Hubei Province.

References

1. Qunlin Zhang, Yijun Mao and Datong Qi. Effect of perforation on the sound transmission through a double-walled cylindrical shell [J]. Journal of Sound and Vibration, 2017, 410: 344-363.
2. Qunlin Zhang. Sound transmission through micro-perforated double-walled cylindrical shells lined with porous material [J]. Journal of Sound and Vibration, 2020:485.
3. Kun Xie, Meixia Chen, Lei Zhang, De Xie. Wave based method for vibration analysis of double-walled cylindrical shells [J]. Applied Acoustics, 2017, 123: 107-122.
4. J.H.LEE & J.KIM. Analysis and measurement of sound transmission through a double-walled cylindrical shell. Journal of Sound and Vibration, 2002, 251(4): 631-649.
5. C. Cacciolati. Sound transmission through a double-walled cylindrical shell with radial elastic links. The Journal of the Acoustical Society of America 87, S74 (1990): 2028353.

Research and Development of Intelligent Cleaning and Mud Content Detection Technology for Machine-Made Sand

Shaoying Jia*, Xiong Zhang and Yelin Lu

Huahao Zhongneng Testing Technology Group Co., Ltd, Beijing 100086, China

Abstract. In this paper, an improved method of sand and mud content detection for concrete is proposed, and a set of sand and mud content detection device for concrete is designed. In this paper, a concrete gas content meter and a bubble parameter analysis system were used to study the effect of sand clay content on the gas content of commercial concrete, the loss rate of gas content in warp time, the bubble spacing coefficient and the distribution of hardened appearance pores. The pore structure test shows that the increase of mud content will appropriately reduce the proportion of harmless pores in machine-made sand concrete, and play a role in refining the pore structure in machine-made sand concrete. By using PLC controller and automatic detection device of liquid level and turbidity, the problems of complicated operation, inconvenient carrying and poor repeatability of test data are solved. It reduces the influence of human operation error and improves the accuracy of test data.

Keywords: sand and gravel; mud content measurement system; concrete; Workability; compressive strength.

1. Introduction

Concrete is composed of cement, sand, stone, water and chemical admixtures, when necessary, formed after cement condensation and hardening, with a certain strength and durability of artificial stone. Sand and stone are one of the main components of concrete, called aggregate, about 70% of the total volume of concrete. Workability refers to the concrete mix is easy to construction operation, can achieve uniform structure, forming dense performance. Workability mainly includes fluidity, cohesiveness and water retention. Mud content is an important index in the quality standard of concrete sand aggregate. High mud content may cause the increase of water demand, affect the workability of concrete mix and the strength of concrete, hinder the full development of cement and aggregate bonding, hinder the normal hydration of cement, or chemical reaction with cement components [1]. There are strict limits on the amount of mud in national and industry standards. Concrete requires that the clay content of natural sand should not be more than 5%, and the clay content of stone should not be more than 1.5%. Some scholars have proposed to use the "washing method in the screen" to detect the mud content of sand. Compared with the standard method, this method has a certain improvement, which can reduce the operation of sand repeatedly pouring back and forth, thus reducing the probability of sand loss. Some scholars have put forward an improved method to solve the problems in the detection method of sand and mud content for concrete. The main method is to screen the sand treated according to the standard method and remove the particles less than 0.075mm that failed to be discharged during the washing process, so as to make the detection result of sand and mud content more accurate. If the above two methods are combined to detect the mud content, its accuracy will be increased, but the possibility of sand loss still exists and the operation is more complicated. This paper designs a set of concrete sand mud content detection device, which can make the sand in the device cleaning, drying and weighing, solve the problem of easy loss of sand, eliminate the influence of human factors. This system is of great significance to control product quality and avoid quality disputes.

2. Machine-made sand mud content detection system

2.1 System Architecture

In this paper, the concrete sand mud content detection device is designed by mixing system, hair dryer system, 1.18mm square hole screen, 0.075mm square hole screen, water container, handle, base and shell. Among them, the stirring system, hair dryer system, 1.18mm square hole screen, 0.075mm square hole screen are located in the shell, the shell is fixed on the base. Considering the need to be weighed together with the device during the test, light materials are preferred for the manufacturing of the device [2]. The total weight should not exceed 15kg. The profile of the device is shown in Figure 1.

* Corresponding author: 34080575@qq.com

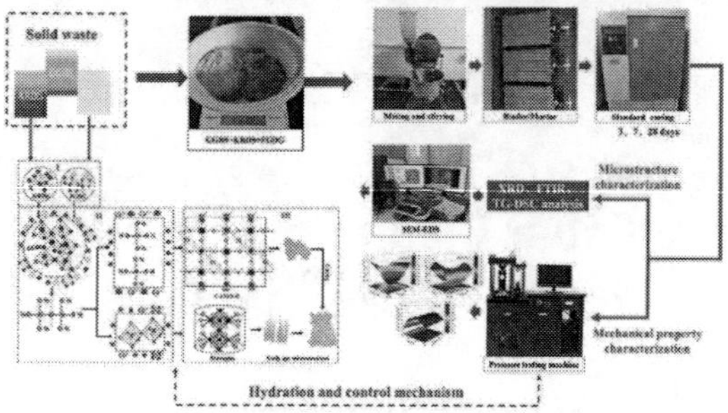

Figure 1. Section diagram of the mud content testing device

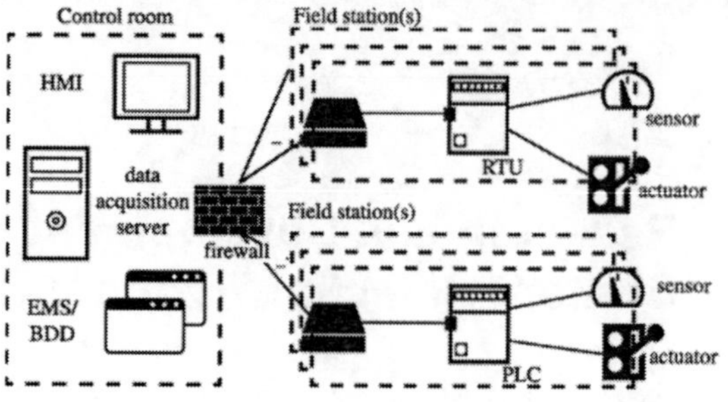

Figure 2. Electronic control system of measuring device

The stirring system is composed of a motor, a rotating shaft and a brush stirring leaf. The motor drives the brush stirring leaves placed on the 1.18mm and 0.075mm screens to rotate in the water through the rotating shaft, so as to achieve the effect of sand cleaning. The hair dryer system consists of two high power hair dryers. After the sand is cleaned, start the hair dryer system, which can make the cleaned sand dry quickly under the condition of stirring and blowing hot air, and at the same time discharge the particles less than 0.075mm suspended in water or failed to discharge during the washing process [3]. One side of the shell of 1.18mm square hole screen and 0.075mm square hole screen are provided with small doors to facilitate the feeding and pouring of materials during the test. In this device, the water container is the supporting equipment when the test device is cleaning the sand. Its diameter is slightly larger than the diameter of the test device, and its height is slightly smaller than the height of the upper layer of the device, so as to prevent the water from being immersed into the upper live system when cleaning the sand.

2.2 Electronic control device

The electronic control system of the measuring device is composed of liquid level sensing device, turbidity detection device, A/D analog-to-digital conversion module, PLC control interface, inlet solenoid relay valve, inlet solenoid relay valve, water spraying device and power module, etc. (Figure 2). PLC transmits the signal of liquid level sensing device and turbidity detection device to the upper computer through the analog-to-digital conversion module [4]. The related configuration software queries the data in the library several times to determine whether the liquid level in and the turbidity in the discharge reach the set upper limit. Then, PLC controller according to the test program written in advance. By controlling the inlet solenoid valve and outlet solenoid valve switch, the mud content test is completed. The controller of the control system itself is a complete set of independent control unit, but also a basic data acquisition node, which can realize the automatic analysis of data and command execution.

2.3 Operation Process

First, open the lid, load 500-1000g sampling sand into the screen sand cylinder, manually start the hot air-drying device of the lid, and begin to dry the sand. After 3h, manually stop the hot air device of the barrel cover and open the balance weighing device of the barrel cover. The balance displays the weight G of sand after drying. PLC controller is energized, open the water inlet pipe valve, 15cm away from the upper part of the screen sand cylinder, cleaning cylinder wall is provided with a level induction device, when the level reaches the set level, induction signal transmission to the PLC controller, PLC controller through the electromagnetic relay valve control closed water [5]. After 120min, the PLC controller controls the outlet electromagnetic relay valve and inlet electromagnetic relay valve to open simultaneously. At this point. Manually start the switch of the precision augmenting electric mixer on the barrel cover, and the sand cleaning process begins. The outlet is connected with an online turbidity detection device. When the effluent turbidity index is lower than the set value, the detection signal is transmitted to the PLC controller, and the PLC controller controls the closing of the inlet electromagnetic relay valve. At the same time, the spray washing device on the inner wall of the fuselage is controlled to start, and the remaining sand on the sand cylinder wall of the screen is washed. At this time, manually turn off the precision augmenting electric mixer. After lmin, the spray washing device will automatically stop working, and all the water in the cleaning cylinder will flow out. The sand cleaning process is over. Manually start the hot air device of the cylinder cover again for sand washing and drying. After 3h, close the hot air device of the barrel cover and open the balance weighing device of the barrel cover. The balance shows the weight of sand G_2 after drying. The quality difference between G_1 and G_2 is the mud-containing quality of the sand sample in this test. The calculation formula of mud content is as follows:

$$Q_a = \frac{G_1 - G_2}{G_1} \times 100\% \tag{1}$$

(kg/m^3) Where Q_a is the mud content, %. G_1 is the mass of the drying sample before the test, g. G_2 is the mass of the drying sample after the test, g. Two samples were taken at the same time for each test. The arithmetic mean value of the two samples was taken as the final test result, accurate to 0.1%.

3. Experimental design

3.1 Raw Materials

The raw materials are 42.5R ordinary Portland cement, sand, stone, water and mud. The mix ratio was C30 concrete, and the same mix ratio was used in the test. Natural sand with different mud content was mixed to carry out the sand mud content test [6]. Sand with mud content of 1.0%, 2.5%, 3.5% and 5% were selected. In the test, all other materials were the same except for the different clay content of sand, and the test mix ratio was shown in Table 1. The sand clay content test is based on "Standard of Sand and Stone Quality and Inspection Method for Ordinary Concrete", and the concrete performance test is based on "Test Method for the Performance of Ordinary Concrete Mix".

Table 1. Concrete test mix (kg/m^3)

Cement	Fly ash	Sand	Stone	Water	Additive limn
290	110	699	1141	160	7.2

3.2 Experimental Results

At present, with the popularization of engineering construction mechanization, the fluidity and slump loss of concrete is more and more high requirements, poor liquidity, slump loss is larger, will make the quality of concrete and construction performance will be seriously affected. As an important index affecting concrete performance, sand clay content should be paid enough attention to. The ratio in Table 2 was used in the test, and the single factor changed the mud content of river sand to 0%, 2.0%, 4.0% and 6.0% respectively. The influence of different mud content of river sand on concrete slump and expansion was investigated, and the test results were shown in Figure 3.

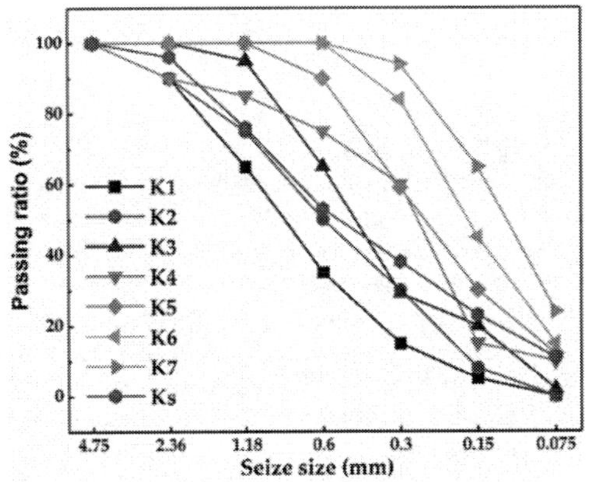

Figure 3. Influence of sand clay content on concrete slump

It can be seen from the experimental results that the expansion degree of each grade concrete decreases with the increase of mud content, and the slump of C40 and C60 concrete also shows the same change trend, but the slump of C25 concrete increases with the increase of mud content within a certain range. Because the amount of cementitious material of C25 concrete is small, the cohesivity and the amount of coating is not good [7]. When the mud content is small, the slump is easy to appear like "straw hat", resulting in small slump and large expansion. When the clay content increases, the cohesiveness of C25 concrete gradually becomes better. This is because mud, as a powdery material, has a certain complement to the cementitious material of concrete, plays the filling role of admixture, and can improve the workability of new mixed concrete. When the mud content increased to a certain amount, the adsorption of polycarboxylic acid admixture increased obviously, and played a leading role, which showed that the concrete fluidity became worse and the loss became faster.

FIG. 4 Experimental results show that, with the increase of mud content, the admixture content of each grade concrete increases continuously when it reaches the corresponding slump, the mud content increases from 2% to 6%, and the admixture content that meets the requirements of fluidity basically doubles about. This is due to the adsorption of clay particles on the admixture, resulting in the reduction of the water reduction effect of the admixture. Under the same conditions, the admixture amount is reduced, so that the concrete cannot reach the expected effect. In addition, with the increase of mud content in the test process, although the concrete can reach the target slump by increasing the admixture content, the expansion degree of concrete mix still decreases. This is because the increase of mud content improves the cohesiveness and water retention of concrete [8]. Although the slump of the mixture can reach, the fluidity becomes worse. The comparison samples without mud showed poor cohesiveness, bleeding and bottom grabbing, and the above phenomena were improved with the increase of mud content, which also indicated that clay had strong adsorption capacity to water and admixtures.

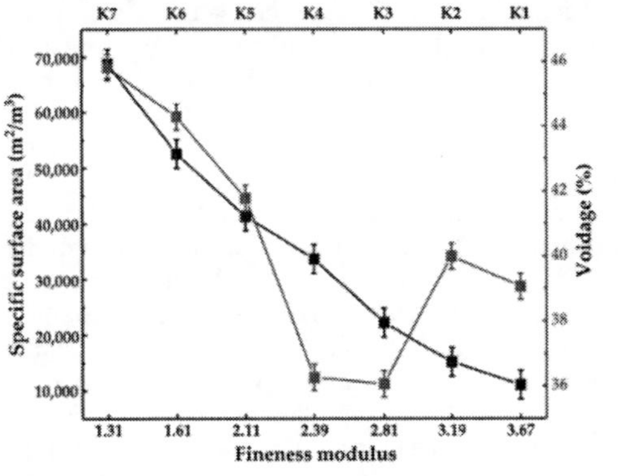

Figure 4. Influence of sand clay content on concrete admixture content

The distribution of pores in concrete hardened appearance is an important component to characterize the appearance quality of concrete, but the causes of the distribution of hardened pores are complex, including mix ratio, admixture and construction technology, etc., will have a great impact on the distribution of pores. From the perspective of concrete performance, there are mainly two influencing factors: The first is the bubble structure inside the concrete. After the plasticity of the concrete is lost, the bubbles close to the formwork wall form pores. The other is the rheological properties of the slurry, the yield stress and plastic viscosity of the slurry directly affect the interface properties between the slurry and the template (figure 5).

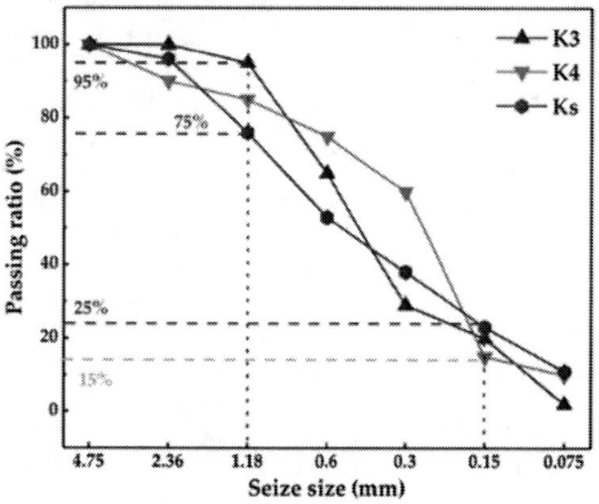

Figure 5. Effect of mechanical sand clay content on pore distribution of concrete hardened appearance

It can be seen from FIG. 5 that with the increase of mud content in machine-made sand, the average size of concrete appearance pores gradually increases, indicating that the increase of mud content has an adverse effect on the appearance bubble structure. However, the increase of plastic viscosity caused by mud is beneficial to the stable adhesion of bubbles to the formwork interface, and the increase of yield stress enhances the stability of the thin wall of the bubble slurry and weakens the agglomeration effect of the stirred bubbles. In general, bubbles with continuous size distribution are formed inside the concrete to adhere to the interface between the formwork and the slurry, thus increasing the average size of bubbles with hardened appearance. The increase of plastic-viscosity of slurry caused by the increase of mud content not only increases the size of pores in appearance, but also increases the number of pores, which is mainly due to the influence of the increase of total gas content. The increase of mud content in machine-made sand not only has a certain ability of auxiliary air entraining for concrete, but also results in the rapid increase of the number of bubbles of all sizes in concrete, and the increase of mud content in an appropriate range can improve the bubble structure of concrete. However, when the mud content increases to more than 4%, the number of pores in concrete hardened appearance gradually decreases with the increase of mud content. In this case, it is not only affected by the decreasing effect of mud on air entraining, but also affected by the continuous increase of the average size of bubbles, that is, the bubble structure inside concrete presents a state of "integration" under certain gas content conditions. The increase in the proportion of large bubbles in all bubbles ultimately leads to a decrease in the number of pores per unit area in the hardened concrete appearance.

4. Conclusion

Whether the clay content of sand used in common concrete mix is accurate or not will have a profound influence on the quality of the final concrete pouring building. The sand and mud content i Bedding device for concrete has solved the disadvantages of complicated operation in conventional test, inconvenient equipment to carry, poor repeatability of test data and too long test operation time. The arrangement of PLC controller, automatic detection device of liquid level and turbidity not only reduces the influence of human operation error. It also improves the accuracy of the test data: the limited space in the barrel is adopted to optimize the spatial structure to the maximum extent, and the portability and flexibility of the concrete sand and mud measurement device is ensured on the premise of ensuring no influence on the test results. Under the same condition of controlling the initial slump of concrete, the gas content of concrete increases first and then decreases with the increase of the mud content of machine-made sand. When the mud content is less than 6%, the gas content increases with the increase of the mud content. When the mud content is more than 6%, the gas content decreases with the increase of mud content.

References

1. Liu Zhaonian, Yue Jiaping, Wang Jiesun, Tingwu, Zhi Qiang. Study on optimization technology of cement slurry system admixture suitable for deep water shallow unconsolidated sandstone formation. Chemistry and Adhesion, vol. 043, pp. 12-16,January 2021.
2. Hu Jianlin, GAO Pengfei, Cui Honghuan, Zhang Yulong. Strength deterioration of soil-cement of iron tailings under dry-wet cycle. Journal of Yangtze River Scientific Research Institute, vol. 39, pp. 95-98,May 2022.
3. Lin Hong-min, Branlan, Peng Jie, et al. Study on the effect of mud content on the fluid curing treatment of sandy soil. Journal of Hebei University of Technology: Natural Science Edition, vol. 39, pp. 65-74,March 2022.
4. Yang Shiwen, Su Aijun, Liu Meng, et al. Influence of mud content and water content on shear strength of granite weathered sand fine aggregate in Three Gorges Reservoir area. Science Technology and Engineering, vol. 20, pp. 89-95,November 2020.
5. Yin Li, Feng Wenjie, Yin Yanshu, et al. Deposition process and model of sandy beach bar under wave action: Based on flume deposition simulation experiment. Acta Sedimentologica Sinica, vol. 40, pp. 13-18,May 2022.
6. Zhao Xingqi, CUI Shoukai, CAI Ya, LU Chong, Yang Xianxian, ZHANG Yong, WANG Weishi, Qing Ping. Metallogenic conditions and prospecting direction of sandstone type uranium in Mahai area, northern margin of Qaidam Basin. Journal of Earth Sciences, vol. 42, pp. 593-604,May 2021.

7. Wang Di, Zhang Yiming, Liu Chuncheng, et al. Inversion and quantitative interpretation of gravity flow sandstone reservoir in deep water area of Baiyun Depression. Computing Techniques for geophysical and geochemical exploration, vol. 43, pp. 79-82,April 2021.
8. Song Yuntao, Dong Guanghui, Xu Xuehong, et al. Experimental study on performance of fine grained concrete based on waste ultra-fine sand. Journal of Yangtze River Scientific Research Institute, vol. 038, pp. 157-161,November 2021.

Dynamic Simulation of Theoretical Mechanics Experiments using Maple

Yanxiang Gong*, Juan Guo, Feng Li

College of Physics and Electronic Engineering, Taishan University, Taian 271000, China.

Abstract. Some experimental phenomena in theoretical mechanics are not suited for realization in an ordinary laboratory. Virtual experiments in theoretical mechanics can be realized using Maple, a scientific computing software. Although most previous studies have only drawn static images of the trajectories of objects, this thesis investigates the use of Maple's animation capabilities to achieve dynamic simulations of theoretical mechanics experiments. The benefits of dynamic simulation are that the trajectory of the object over time can be seen and the motion characteristics of the object, such as the direction of approach of the incoming track, can be fully understood. The dynamic simulation of theoretical mechanics experiments will fully mobilize students' interest in learning, transform abstract physical models into visual physical images and deepen their understanding of physical phenomena. In addition, virtual experiments enable students to train their programming skills and improve their scientific research literacy.

Keywords: Maple, theoretical mechanics, educational technology, animation simulation.

1. Introduction

Physics experiments are an important part of physics teaching and one of the three main research tools in physics (the other two are theoretical analysis and numerical simulation), but some experiments are not suitable for implementation in ordinary laboratories, such as the Foucault pendulum experiment in theoretical mechanics and the perihelion inlet phenomenon of planets. As a result, the understanding of abstract physical problems becomes a bottleneck for learning, which affects the learning effect, due that students often only know the analytical derivation process for some mechanical phenomena, but not the physical images. Describing the physical image of a theoretical mechanics problem vividly with virtual experiments is the best way to learn theoretical mechanics. Rather, it inspires our interest in learning and deepens our understanding of physical theory, while allowing us to learn about simple programming, drawing and data processing [1, 2]. Lots of previous studies introducing virtual experiments made static images for mechanics experiments, however. It leads to students knowing the trajectory of the object, but not being able to see dynamically and vividly the trajectory of the object over time. Dynamic images will be able to give more information, for example, the incoming direction of the object's trajectory, its instantaneous position, etc.

It is also one of the issues to be considered in depth when choosing the right programming language for a virtual experiment. It is difficult to master traditional scientific computing languages such as C and Fortran, which involve a lot of repetitive and complex programming work and are therefore not suitable for undergraduate students. More advanced and powerful computational languages such as Maple, Matlab, Mathematica, and Mathcad are common choices [3, 4]. Compared to other software, Maple selected has the following advantages: First, the programming is easy to learn and simple, which is incomparable to the usual computing languages. C and Fortran can only be used for numerical operations, and Matlab is not as specialized as Maple, although it can do symbolic operations, for example, symbolic variables or symbolic expressions must be defined in advance. Matlab's symbolic operations start with the Maple kernel, whose function names are obvious and easy to remember. It can compute huge integers exactly. Secondly, there is a high degree of integration of functions. It combines image processing, symbolic operations, and numerical calculations in one. It is something that is not available in ordinary high-level languages. Theoretical mechanical problems such as the perihelion motion of Mercury and the scattering of alpha particles can be easily simulated with Maple, which offers powerful drawing capabilities, including 1D, 2D and 3D image drawing. Particularly, dynamics simulation of mechanical problems can be easily achieved by Maple's animation and various useful drawing packages. In this thesis, two typical theoretical mechanics problems are used as examples to study how to

* Corresponding author: yxgong@sina.com

realize "dynamic" simulations of theoretical mechanics experiments using the Maple language.

2. Dynamic Simulation of Mercury's Perihelion Approach

One of the important experimental verifications of Mercury's perihelion progression for general relativity. In case only the Newtonian gravitational interaction between the Sun and Mercury is considered (both as masses), Mercury's orbit is a closed ellipse. But progression will occur in its orbit if relativistic corrections are considered. In the polar coordinate system, the differential equation for Mercury's orbit is [5].

$$\frac{d^2u}{d\theta^2}+u=\frac{Gm_0}{h^2}+\frac{3Gm_0}{c^2}u^2 \quad (1)$$

Where $u=1/r$, G is the universal gravitational constant, m_0 is the mass of the Sun, h is twice the speed of Mercury's grazing surface, and c the speed of light in a vacuum. After parameters: $a=Gm_0/h^2$, $b=3Gm_0/c^2$ are introduced, the equation can be rewritten as

$$\frac{d^2u}{d\theta^2}=-u+a+bu^2 \quad (2)$$

Let $x=u$, $y=du/d\theta$, rewrite the above second-order differential equation as a system of first-order differential equations, i.e.

$$\begin{cases}\dfrac{dx}{d\theta}=y\\ \dfrac{dy}{d\theta}=-x+a+bx^2\end{cases} \quad (3)$$

The parameters need to be chosen appropriately to exaggerate the orbital procession due to the small actual correction term when solving numerically. A result is shown in Figure 1.

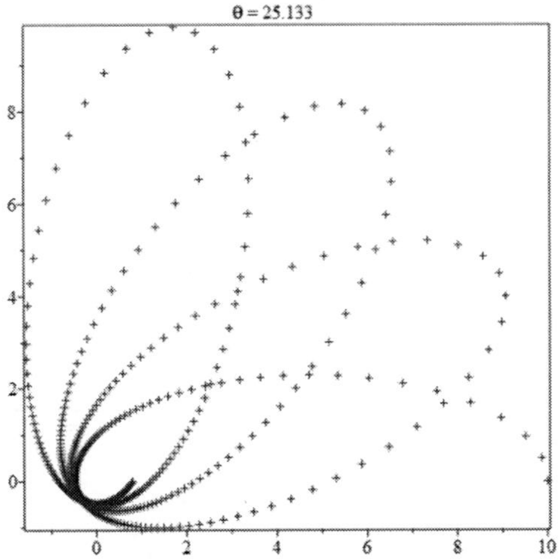

Figure 1. Dynamic Trajectories of the Perihelion Approach of the Mercury, with the cross marks varying with angle (θ as a function of time)

The procedure, implemented in Maple, is as follows:

```
restart;
with(plots);
a := 1;
b := 0.01;
eq := diff(u1(theta), theta) - u2(theta) = 0,
diff(u2(theta), theta) + u1(theta) - a - b*u1(theta)^2 = 0;
sol := dsolve({eq, u1(0) = 0.1, u2(0) = 0},
{u1(theta), u2(theta)}, numeric);
Mercury := proc(z)
local U1, X, Y, note;
global theta, sol;
U1 := eval(u1(theta), sol(z));
X := 1.*cos(z)/U1;
Y := 1.*sin(z)/U1;
plot([[X, Y]], style = point, axes = boxed, color = blue);
end proc;
animate (Mercury, [theta], theta = 0. 8*Pi, frames = 500,
trace = 500);
```

3. Dynamic Simulation of Foucault Pendulum Motion

The Foucault pendulum, a device to show the rotation of the Earth, was first made in Paris in 1851 by the French physicist Foucault. The special trajectory of this pendulum is due to the rotation of the Earth and the influence of the pendulum by the Coriolis force. Its physical model is as follows [6]. The coordinate system is shown in Figure 2.

$$\ddot{x}=2m\omega\dot{y}\sin\lambda-p^2x \quad (4)$$

$$\ddot{y}=-2m\omega\dot{x}\sin\lambda-p^2y \quad (5)$$

where m is the mass of the Foucault pendulum, x˙, y˙ are the velocity components in the Cartesian coordinate system, x¨, y¨ are the acceleration components, ω is the angular velocity of the Earth's rotation, and λ is the dimensional angle of the sphere where the Foucault pendulum is located. p^2 = g/l, is the acceleration of gravity and is the pendulum length of the Foucault pendulum. A simulation result is shown in Figure 3.

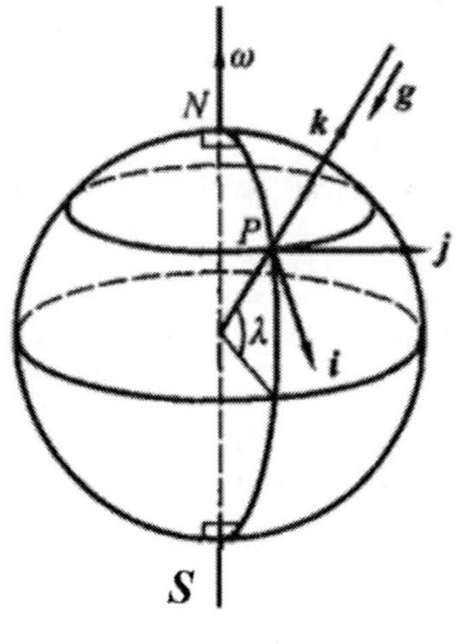

Figure 2. Schematic Diagram of the Earth's Coordinate System [6]

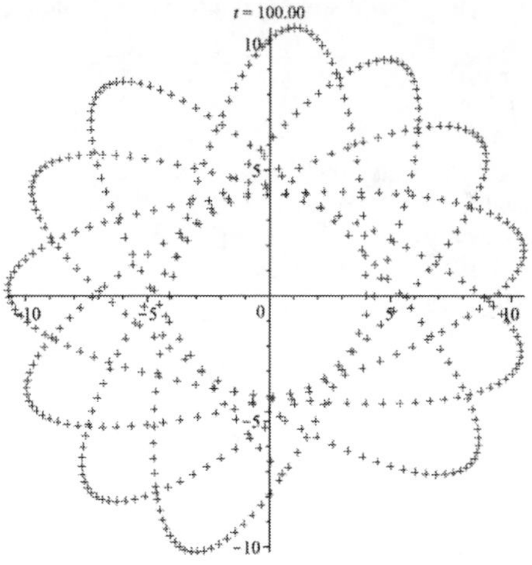

Figure 3. Dynamic Trajectories of the Foucault Pendulum, with the cross markers changing dynamically with time

The procedure, implemented in Maple, is as follows:

```
restart;
with(plots);
lambda := (60*Pi)/180;
g := 9.8;
l := 67;
omega := 0.03;
sys := diff(x(t), t $ 2) - 2*omega*diff(y(t), t)*sin(lambda) + g*x(t)/l = 0, diff(y(t), t $ 2) + 2*omega*diff(x(t), t)*sin(lambda) + g*y(t)/l = 0;
sol := dsolve({sys, x(0) = 4, y(0) = 0, D(x)(0) = 0, D(y)(0) = 4}, {x(t), y(t)}, numeric);
drawFuke := proc(z)
local X1, Y1;
global x, y, sol;
X1 := eval(x(t), sol(z));
Y1 := eval(y(t), sol(z));
plot([[X1, Y1]], style = point, scaling = constrained, color = red);
end proc;
animate (drawFuke, [t], t = 0 .. 100, frames = 500, trace = 500);
```

4. Summary

The analysis in this thesis shows that dynamic experiments in theoretical mechanics can be easily produced using Maple's animation capabilities and the rich package of numerical calculation functions. It is specifically summarized as follows. First, the kinematic characteristics of the Foucault pendulum and Mercury's perihelion inlet can be dynamically displayed using Maple's solved differential equations and animation functions, making it easier for students to understand their physical images. Second, information such as the direction of the orbital approach and the instantaneous position of the object can be understood through dynamic maps; the influence of parameters such as initial conditions and dimensions on dynamic trajectory maps can be studied to gain a deeper understanding of physical phenomena. Third, Maple software is very suitable for experimental simulation of theoretical mechanics. Thus, it enables university students to cultivate an interest in learning theoretical physics, acquire programming knowledge, conduct early research training, etc.

References

1. Dong Shaoguang. Teaching research on classical physics complex problems based on MATLAB visualization program. Chinese Physics Bulletin, 2021 (09): 22-27.
2. Zhou Qunyi, Mo Yunfei, Zhou Lili, Hou Zhaoyang. The application of MATLAB in quantum mechanics learning. College physics, 2021 (40): 06.
3. Yu Shixing. Application of Maple in electromagnetic field and Electromagnetic wave theory teaching. Education and teaching forum, 2019, 34: 187-188.
4. Yang Xiaobo. Application of Maple software in the teaching of definite integral in advanced Mathematics. Education modernization, 2020, 7 (20): 1-4.
5. Peng Fanglin. Computer Simulation of Theoretical Mechanics[M]. Tsinghua University Press, 2002, 120-121.
6. Zhou Yanbo. Theoretical Mechanics Course (Second Edition) [M]. Beijing: Higher Education Press, 2018, 178-179.

https://doi.org/10.1051/matecconf/202338001022

Influence of external load on forced vibration of rectangular plate

Peng Wang [1], Hao Zhang [1], Qingyong Niu [2], Xiaoming Cao [1], Min Liu [1,*]

[1] Wuhan Second Ship Design and Research Institute, Wuhan 430205, China
[2] College of Science, Wuhan University of Science and Technology, Wuhan, 430065, China

Abstract. Rectangular plate is one typical structure with wild application in engineering, and the force vibration is the basis for analysis of sound radiation of plate. In this work, numerical simulation is conducted for force vibration of rectangular plate, and the influence of external load is also studied. Numerical results show that, the rectangular plate will resonate under external load at certain driven frequency, and it's strictly related to structural modal frequencies. Structural vibration response will change obvious with the change of location of external load. When the location is fixed, with the increase of action region of external load, the acceleration of plate will decrease and some new resonance frequencies will be observed. Conclusions here will lay foundation for further work on sound radiation of plate.

1. Introduction

Rectangular plate is one typical structure which has wildly application in aero-space, ship structure, marine structures, and so on. Vibration characteristic of rectangular plate receives great attentions from researchers, since it's the foundation for sound radiation of plate.

Bardell et.al.[1] applied the Ritz method to analyze the in-plane vibration of rectangular plate with different boundary conditions. Gorman [2-6] put forward a new superposition method for solving the vibration of rectangular plate with free, simply-supported and clamped boundary conditions. Based on the differential quadrature method (DQM), Xing et.al.[7] put forward a new method with high accuracy for in-plane vibration of rectangular plate. Du et.al.[8-9] use the Fourier series to analyze the vibration of rectangular plate with elastic boundary conditions. But, it should also be important to get clear the influence of external load on structural vibration.

In this work, numerical analysis for forced vibration of rectangular plate is carried out, and the influence of external load (i.e., location and action region) on structural vibration is also conducted.

2. Numerical simulation of forced vibration

2.1 Structural model

The finite element model for rectangular plate is shown in Fig.1, the plate is of length 2640mm, width 800mm, thickness 10mm. Both long edges are clamped. Material for the plate is 316L steel.

Fig. 1 model of rectangular plate

* Corresponding author: 549931866@qq.com

2.2 Forced vibration

Harmonic vibration of plate is conducted here by applying an external point unit force at the center of plate in normal direction. The frequency range for numerical analysis is 10Hz-3000Hz. Mean acceleration of plate is plotted in Fig.2.

It's note that the plate will resonate at certain excitation frequency, i.e., 80Hz, 430Hz, 1120Hz, and so on. Due to the setting of step length of 10mm in numerical simulation, little difference is note between structural modal frequency and resonance frequency.

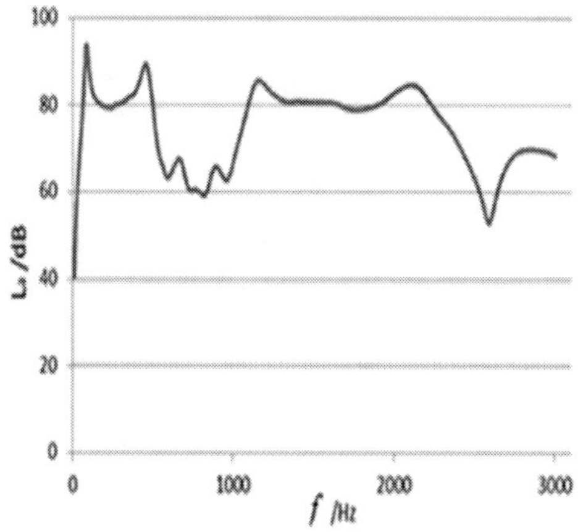

Fig. 2 structural acceleration distribution versus driven frequency

3. Influence of external load

3.1 Location of point excitation

To analyze the influence of load location on structural vibration response, three cases of location at the center, L/4 and short edge are calculated, and the mean acceleration of plate is plotted in Fig.3.

It's noted that, the change of load location will affect the amplitude of vibration acceleration, resonance frequency remains the same, and new resonance frequency will be observed. Since the structural modal frequencies do not change. But the change of load location will lead to different contribution of structural mode to vibration response, and finally result in different structural responses.

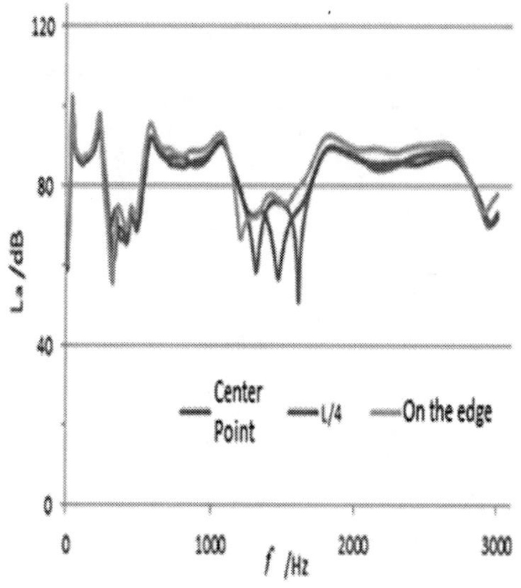

Fig. 3 Distribution of structural acceleration versus driven frequency in case of different load location

3.2 Action region of external load

Point load is ideal, and action region should be considered in engineering. Therefore, the effect of action region on structural vibration response is studied in this section. Three cases of action region, 120 mm×70 mm, 120 mm×100 mm, 340 mm × 180mm, ware calculated and compared. Numerical results are shown in Fig.4. The total acceleration level in 10Hz-3000Hz are listed in table 1. It's noted that: 1) general distributions of acceleration versus driven frequency from three cases are similar, main resonant frequencies are the same since structural modal frequency do not change. 2) With the increase of action region, few new resonance frequencies will be noted, since the change of action region will change the contribution of structural modes to vibrational response, and few modal contribution will increase and be the main part; 3) With the increase of action region, structural acceleration level will decrease.

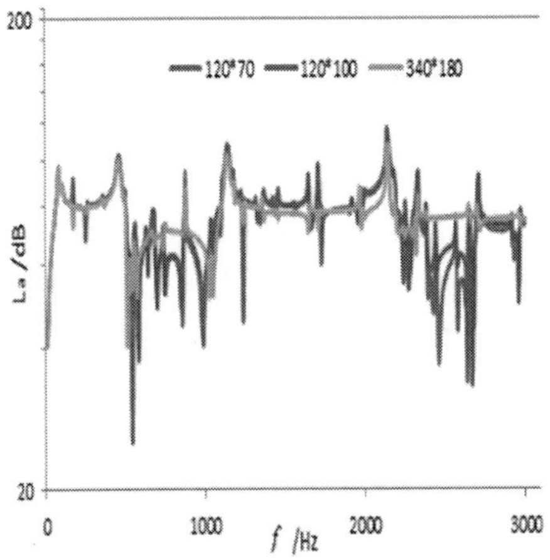

Fig. 4 distribution of acceleration versus driven frequency with different action regions

Table 1. Comparative analysis of total acceleration in case of different action region

Frequency range	120mm×70 mm	120mm×100 mm	340mm×170 mm
10Hz-3000Hz	108.85 dB	106.67 dB	102.70 dB

4. Conclusion

In this work, numerical simulation is conducted for force vibration of rectangular plate, and the influence of external load is also studied. Numerical results show that, the rectangular plate will resonate under external load at certain driven frequency, and it's strictly related to structural modal frequencies. Structural vibration response will change obvious with the change of location of external load. When the location is fixed, with the increase of action region of external load, the acceleration of plate will decrease and some new resonance frequencies will be observed.

Conclusions here will help to give a better understanding of the sound radiation of rectangular plate.

Acknowledgements

We sincerely thank the financial support from the Hubei Provincial Natural Science Foundation (No. ZRMS2020002231) and the Young Top-notch Talent Cultivation Program of Hubei Province.

References

1. Bardell N S, Langley R S, Dunsdon J M. On the free in-plane vibration of isotropic rectangular plates[J]. Journal of Sound and Vibration, 1996, 191(3): 459-467.
2. Gorman D J. Accurate analytical type solutions for the free in-plane vibration of clamped and simply supported rectangular plates[J]. Journal of Sound and Vibration, 2004, 276(1): 311-333.
3. Gorman D J. Exact solutions for the free in-plane vibration of rectangular plates with two opposite edges simply supported[J]. Journal of Sound and Vibration, 2006, 294(1-2): 131-161.
4. Gorman D J. Free in-plane vibration analysis of rectangular plates with elastic support normal to the boundaries[J]. Journal of Sound and Vibration, 2005, 285(4-5): 941-966.
5. Gorman D J. Free in-plane vibration analysis of rectangular plates by the method of superposition[J]. Journal of Sound and Vibration, 2004, 272(3): 831-851.
6. Xing Y F, Liu B. Exact solutions for the free in-plane vibrations of rectangular plates[J]. International Journal of Mechanical Sciences, 2009, 51(3): 246-255.
7. Du J T, Liu Z G, Li W L, et al. Free in-plane vibration analysis of rectangular plates with elastically point-

supported edges[J]. Journal of Vibration and Acoustics-Transactions of the ASME, 2010, 132(3).

8. Du J T, Li W L, Jin G Y, et al. An analytical method for the in-plane vibration analysis of rectangular plates with elastically restrained edges[J]. Journal of Sound and Vibration, 2007, 306(3-5): 908-927.

Robust Attitude Control of Helicopters with Three-Degree Freedom Based on Intermittent Event-Triggered Mechanism

Chaoyang Zhang *

School of Mechanical and Materials Engineering, North China University of Technology, China

Abstract. In the field of the widely used helicopter, this paper studies whether the helicopters with three-degree freedom (3DOF helicopters) have a new control method of stable attitude. Therefore, two research methods are adopted. One is to use intermittent control, so as to achieve control stability in a short time without its inputting. The other is to find and update the time needed to meet the next control through the event-triggered mechanism. A theorem is given to prove the feasibility of these methods, and Lyapunov's second law is also applied in this paper to analyze the stability of the model designed for robust attitude control. Besides, we use the linear matrix inequality solver to solve the feedback gain of the controller in the theorem. Finally, through MATLAB simulation, the system error state of helicopters tends to zero under intermittent control, which highlights the advantages of short communication time and few times of simulation, thus achieving the simulation goal.

Key words: Intermittent Control, Event-Triggered Mechanism, Helicopter Homeostatic Control, LMI Toolbox

1. Introduction

Helicopter has been a very popular flight equipment recently [1] because it can be operated remotely with a wide range of applications in aerial photography, agriculture, reconnaissance, surveying, mapping, disaster relief, etc [8]. Thus, it is necessary to study helicopters. However, we have observed that the influence of various factors during aerial work prevents helicopters from keeping a relatively stable attitude for tasks at all times [2], which makes it much more difficult to perform tasks. Therefore, it is also imperative to study its attitude stability control.

Nowadays, there are many methods of helicopter attitude control. For example, the robust attitude control with wind resistance disturbance and information transmission delay is introduced [3] by comparing its efficiency and stability with the traditional H-infinity control model; an adaptive model based on fault-tolerant control is designed for attitude control so that helicopters can work in a larger controllable range [4] [5]. The helicopter attitude based on neural network technology is studied [6] [9], from which the uncertainty and external disturbance of the reconstructed model is approached to realize the attitude tracking control of 3DOF helicopters. A modified adaptive law is proposed to approximate the upper bound of disturbance by using the attitude tracking control model so that the tracking error converges to a preset region with a small overshoot in the user-defined time [7]. Rigatos, G. who discussed the optimal model of nonlinear control realized the fast and accurate tracking of all state variables of three-dimensional helicopters under the moderate change of control input [10]. A robust unknown input estimation scheme based on a second-order sliding mode observer is put forward to ensure the finite-time convergence of the boundary layer [11] [19]. A control model based on quaternion multi-body coupling modeling is studied to improve the workspace and application range of micro four-axis aircraft [12]. Besides, the robust control model of multi-mode switching control is discussed [13] [14]. By introducing the concept of PI control, the pitch and stroke angles can be tracked with less error. The possibility of helicopter control by mechanism search method is applied [15]. After ESC is applied to PID controller tuning, it is obtained that the pitch angle and its tracking error converge to the vicinity of the expected working point, which has high robustness. Step-by-step control of three-dimensional helicopters model with input and output constraints adopts enhanced state observer to estimate an unknown state, unexpected external disturbance, and modeling uncertainty [16]. The application of a dual-loop control system of a laboratory helicopter with 3DOF pitch, elevation, and flight is studied [18]. Similar examples can also be found in structural control, machine learning, and so on. In order to ensure the control performance of helicopter homeostatic systems in practical application, the widely used robust control as a technology is used to deal with system uncertainties and external disturbances. In this paper, the methods of intermittent control and event-

* Corresponding author: crysknife07@mail.ncut.edu.cn

triggered control are applied, making the following main contributions.

- Compared with the traditional control method, the actuator is intermittent control, which saves control time and cost, realizing intermittent communication.
- After the event trigger condition is adopted, the data transmitted by the signal in the network will be reduced. Only when the signal meets the trigger condition can it be output to the control component, which further reduces the communication burden in the network and realizes the stability of control with less data flow.

2. Modeling and Controller Design

2.1 Modeling

As for a common 3DOF helicopter, the pitch and tilt angle can be defined by its motion attitude, so as to determine its error system model. This 3DOF dynamic error control model is discussed in an article by Xiaoyuan Zhu [5]. The tilt and pitch angles are defined as ε_r and p_r, with its error system model defined as

$$\dot{s}(t) = Ms(t) + N\sigma(t) + Wf(t) \qquad (1)$$
$$s(t) = \left[\varepsilon_r, \dot{\varepsilon}_r, \ddot{\varepsilon}_r, p_r, \dot{p}_r, \ddot{p}_r\right]^T$$

M, N, W is the known matrix with appropriate dimensions. Meanwhile, $s(t) \in R^n$ describes the state of the system, $\sigma(t) \in R^m$ represents the control signal input, $f(t)$ indicates the external disturbance, and M, N, W is obtained from the known physical model. For the set model shown in the following figure, we can get a result after listing the relationship between the tilt angle and the pitch angle.

$$M = \begin{bmatrix} 0 & 1 & 0 & 0 & 0 & 0 \\ 0 & 0 & 0 & 0 & 0 & 0 \\ 1 & 0 & 0 & 0 & 0 & 0 \\ 0 & 0 & 0 & 0 & 1 & 0 \\ 0 & 0 & 0 & 0 & 0 & 0 \\ 0 & 0 & 0 & 1 & 0 & 0 \end{bmatrix}, \quad N = \begin{bmatrix} 0 & 0 \\ \frac{K_f L_a}{J_\varepsilon} & \frac{K_f L_a}{J_\varepsilon} \\ 0 & 0 \\ 0 & 0 \\ \frac{K_f L_h}{J_p} & \frac{K_f L_h}{J_p} \\ 0 & 0 \end{bmatrix}, W = \begin{bmatrix} 0 & 0 & 0 & 0 \\ 1 & -1 & 0 & 0 \\ 0 & 0 & 0 & 0 \\ 0 & 0 & 0 & 0 \\ 0 & 0 & 1 & -1 \\ 0 & 0 & 0 & 0 \end{bmatrix}$$

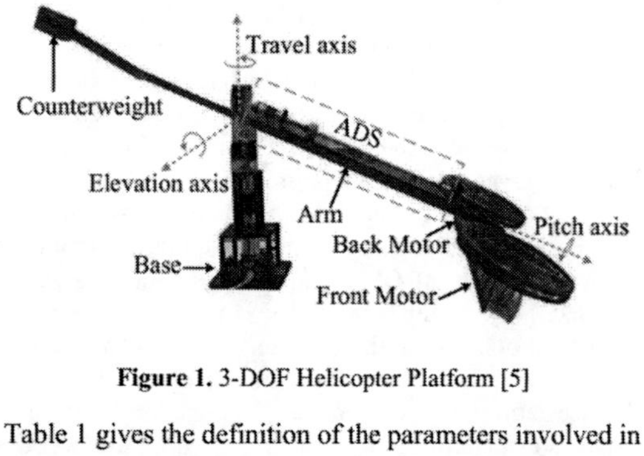

Figure 1. 3-DOF Helicopter Platform [5]

Table 1 gives the definition of the parameters involved in the above-mentioned matrix M, N, W.

Table 1. Symbol Specification1

Symbol	Definition
K_f	Propeller force-thrust constant
L_h	Distance from pitch axis to either motor
L_a	Distance from helicopter body to elevation axis
J_ε	Moments of inertia about elevation axis
J_p	Moments of inertia about pitch axis

As for this dynamic error model, after determining its basic parameters, it needs to be controlled to stabilize the error level. Therefore, the system is designed with intermittent control.

2.2 Controller Design

Based on the intermittent control method, the control input is defined. We assume as followed:

$$\sigma(t) = \begin{cases} Ks(t_k), & t_k \leqslant t < t_k + \eta_k \\ 0, & t_k + \eta_k \leqslant t < t_{k+1} \end{cases} \qquad (2)$$

η_k is randomly generated between $\left[\underline{\eta}, \overline{\eta}\right]$. This method adopts a segmented way, that is, a control input is generated to the controller when $t = [t_k, t_k + \eta_k)$, so that the helicopter does not lose balance or take control when $t = [t_k + \eta_k, t_{k+1})$. On the basis of the controller, we can determine a new system model with intermittent control method, which is

$$\dot{s}(t) = \begin{cases} Ms(t) + NKs(t_k) + Wf(t), & t \in [t_k, t_k + \eta_k) \\ Ms(t) + Wf(t), & t \in [t_k + \eta_k, t_{k+1}) \end{cases} \qquad (3)$$

We continue to make the actuator intermittently control at a specific time by setting the trigger mechanism, clarifying the limit conditions of the control time, and updating the trigger conditions the next time. The trigger is set to:

$$\begin{cases} t_{k+1} = inf\{t > t_k + \eta_k \mid \varphi(t) \geqslant 0\} \\ \varphi(t) = e^T(t)\Omega e(t) - \alpha(t)s^T(t)\Omega s(t) \\ e(t) = s(t) - s(t_k) \\ \alpha(t) = \alpha_1 - (\alpha_1 - \alpha_2)\tanh\left(\dfrac{\alpha_3 e^T(t)\Omega e(t)}{s^T(t)\Omega s(t)}\right)e^{-\|e(t)\|} \end{cases} \qquad (4)$$

As for the system after setting the controller, the control process is better expressed in the form of a flow chart. As shown in Figure 2, the sensor senses the unstable backward trigger of the attitude power system to seek control. The trigger updates the control time, with the controller and the zero-order holder passing the processing before issuing the instruction to the actuator. Finally, the actuator controls and feeds back the actual result to the attitude power system.

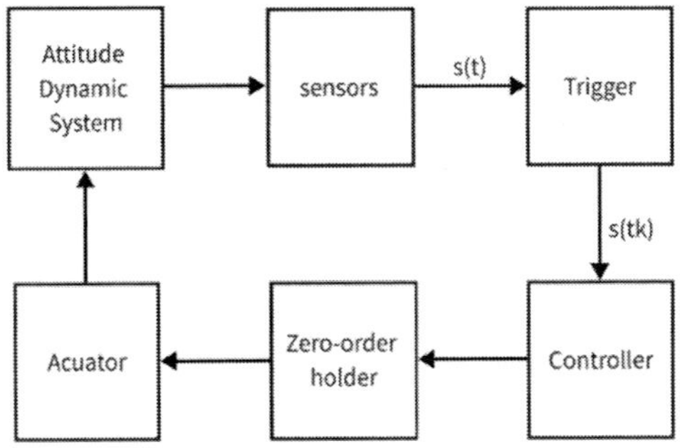

Figure 2. System Control Flow Chart

3. Main Results

In this paper, the following theorem is given to prove that the system error model can achieve exponential stability. Before the main results are given, the following hypothesis is necessary. Under the designed control scheme, this hypothesis is used to restrict the update time interval to fluctuate within a relatively stable interval.

Assumption: As for $t \in [0, +\infty)$, the positive numbers C_1, C_2, C_3, C_4 exist to reach the following result:

$$C_3 \geq sup\{t_{k+1} - t_k - \eta_k\}, \quad C(t) \leq C_1 + \frac{t}{C_2}, \quad C_4 \leq sup\{\eta_k\} \quad (5)$$

Besides, $C(t) = card\{k \in N | t > t_k + \eta_k\}$

Theorem 1: For any parameter α_1, α_2, the closed-loop error system (1) can be exponentially stable and has index γ with the performance of H_∞, if there are any symmetric matrices T and positive definite matrices P_i, R_i, Q_i to make the following inequality $\dot{V}_i + 2(-1)^{i-1} V_i(t) \leq 0$ tenable.

$$\begin{bmatrix} 2\lambda_1 P_1 + Q_1 - \eta_k R_1 + 2(\alpha_1 - \alpha_2)\Omega & P_1 + \frac{G_1 M}{2} & \eta_k R_1 & \frac{G_1 M}{2} \\ * & \eta_k^2 R_1 - G_1 & 0 & \frac{-G_1 + G_1 NK}{2} \\ * & * & -\eta_k R_1 - e^{-2\lambda_1\eta_k} Q_1 & 0 \\ * & * & * & -\Omega \end{bmatrix} < 0 \quad (6)$$

$$\begin{bmatrix} Q_2 - 2\lambda_2 P_2 - \eta_k R_2 & P_2 + \frac{G_3 M}{2} & \eta_k R_2 \\ * & \eta_k^2 R_2 - G_3 & 0 \\ * & * & -\eta_k R_2 - e^{2\lambda_2\eta_k} Q_2 \end{bmatrix} < 0 \quad (7)$$

Proof: As for this closed-loop system, all states in the error dynamics model are selected to evaluate the performance of the robust controller.

$$z = Cs, C = I_6$$

Performance metric is $\|z\|_2 < \gamma \|f\|_2$ and so is γ.

According to Lyapunov's second law, the following $V(t)$ is selected.

$$V(t) = \begin{cases} V_1(t), & t_k \leq t < t_k + \eta_k \\ V_2(t), & t_k + \eta_k \leq t < t_{k+1} \end{cases} \quad (8)$$

$$V_i(t) = s^T(t) P_i s(t) + \int_{t-\eta_k}^{t} e^{2(-1)^i \lambda_i (t-u)} s^T(u) Q_i s(u) du + \eta^* \int_{-\eta_k}^{0} \int_{t+\theta}^{t} e^{2(-1)^i \lambda_i (t-u)} \dot{s}^T(u) R_i \dot{s}(u) du d\theta \quad (9)$$

and $i = 1, 2$

To assume $V_{i1} = s^T(t) P_i s(t)$ (10)

$$V_{i2} = \int_{t-\eta_k}^{t} e^{2(-1)^i \lambda_i (t-u)} s^T(u) Q_i s(u) du \quad (11)$$

$$V_{i3} = \eta_k \int_{-\eta_k}^{0} \int_{t+\theta}^{t} e^{2(-1)^i \lambda_i (t-u)} \dot{s}^T(u) R_i \dot{s}(u) du d\theta \quad (12)$$

We can get the the following from (12)(9)(10)(11)(12)

$$V(t) = V_{i1}(t) + V_{i2}(t) + V_{i3}(t) \quad (13)$$

As for V_{i1}, V_{i2}, V_{i3}, derivations are gained respectively:

$$\dot{V}_{i1}(t) = s^T(t) P_i \dot{s}(t) + s^T(t) P_i \dot{s}(t) = 2 s^T(t) P_i \dot{s}(t) \quad (14)$$

$$\dot{V}_{i2}(t) = s^T(t) Q_i s(t) - e^{2(-1)^i \lambda_i \eta_k} s^T(t-\eta_k) Q_i s(t-\eta_k) + 2(-1)^i \lambda_i V_{i2}(t) \quad (15)$$

$$\dot{V}_{i3}(t) = \eta_k^2 \dot{s}^T(t) R_i s(t) - \eta_k \int_{-\eta_k}^{0} e^{-2(-1)^i \lambda_i \theta} \dot{s}^T(t+\theta) R_i \dot{s}(t+\theta) d\theta + 2(-1)^i \lambda_i V_{i3}(t) \quad (16)$$

As for the integral term in V_{i3}, the critical value can be estimated by scaling method because it can not be solved concretely.

We have $\Phi(\mu) = -\eta_k \int_{-\eta_k}^{0} e^{-2(-1)^i \lambda_i \theta} \dot{s}^T(t+\theta) R_i \dot{s}(t+\theta) d\theta$ and assume $\mu = t + \theta$.

$$\therefore \Phi(\mu) = -\eta_k \int_{t-\eta_k}^{t} e^{-2(-1)^i \lambda_i (\mu - t)} \dot{s}^T(\mu) R_i \dot{s}(\mu) d\mu$$

We also have $Z(\mu) = \begin{cases} e^{2\lambda_i(\mu - t)}, & i = 1 \\ e^{-2\lambda_i(\mu - t)}, & i = 2 \end{cases}$

$$\therefore \begin{cases} Z(\mu) \leq Z(t-\eta_k), \Phi_{max} = -\eta_k e^{-2\lambda_i \eta_k} \int_{t-\eta_k}^{t} \dot{s}^T(\mu) R_i \dot{s}(\mu) d\mu & ,i = 1 \\ Z(\mu) \leq Z(t) & ,\Phi_{max} = -\eta_k \int_{t-\eta_k}^{t} \dot{s}^T(\mu) R_i \dot{s}(\mu) d\mu & ,i = 2 \end{cases} \quad (17)$$

It is observed that there is an integral term $\int_{t-\eta_k}^{t} \dot{s}^T(\mu) R_i \dot{s}(\mu) d\mu$ in both cases of $\Omega(\mu)$ in (17) and the results as followed using jasen inequality for this term:

$$\int_{t-\eta_k}^{t} \dot{s}^T(\mu) R_i \dot{s}(\mu) d\mu \leq -\left(\int_{t-\eta_k}^{t} \dot{s}(\mu) d\mu\right)^T R_i \left(\int_{t-\eta_k}^{t} \dot{s}(\mu) d\mu\right) = -[s(t) - s(t-\eta_k)]^T R_i [s(t) - s(t-\eta_k)] \quad (18)$$

After the inequality (18) is brought in (17) to eliminate the integral term, the following can be obtained.

$$\begin{aligned} \dot{V}_i(t) &= \dot{V}_{i1} + \dot{V}_{i2} + \dot{V}_{i3} \\ &\leq 2s^T(t) P_i \dot{s}(t) + s^T(t) Q_i s(t) - e^{2(-1)^i \lambda_i \eta_k} s^T(t-\eta_k) Q_i s(t-\eta_k) + 2(-1)^i V_{i2} \\ &\quad + \eta_k^2 \dot{s}^T(t) R_i \dot{s}(t) + 2(-1)^i \lambda_i V_{i3} \\ &\quad - [s(t) - s(t-\eta_k)]^T R_i [s(t) - s(t-\eta_k)] \begin{cases} Z(t-\eta_k) & ,i = 1 \\ Z(t) & ,i = 2 \end{cases} \end{aligned} \quad (19)$$

When $i = 1$, (17) will be brought in the equation (19) about V_1

$$\begin{aligned}\dot{V}_1(t)+2\lambda_1V_1(t)=&2s^T(t)P_1\dot{s}(t)+s^T(t)Q_1s(t)-e^{-2\lambda_1\eta_k}s^T(t-\eta_k)Q_1s(t-\eta_k)\\&+\eta_k^2\dot{s}^T(t)R_1\dot{s}(t)+2\lambda_1s^T(t)P_1s(t)\\&-\eta_k[s^T(t)R_1s(t)-s^T(t)R_1s(t-\eta_k)]\\&-\eta_k[-s^T(t)R_1s(t-\eta_k)+s^T(t-\eta_k)R_1s(t-\eta_k)]\end{aligned}\tag{20}$$

Because $\dot{s}=Ms(t)+N\sigma(t)+Wf(t)$, so the following is obtained.

$$0=[\dot{s}^T(t)\times G_1+e^T(t)\times G_2]\times[Ms(t)+N\sigma(t)+Wf(t)-\dot{s}(t)]\tag{21}$$

G_1,G_2 is a free matrix with known dimensions, which is used to simplify the difficulty in the solution.

$s(t_k)=s(t)-e(t)$ is brought into (20) to construct the item of $e^T(t)\times e(t)$

$$\begin{aligned}\dot{V}_1(t)+2\lambda_1V(t)=&2s^T(t)P_1\dot{s}(t)+s^T(t)Q_1s(t)-e^{-2\lambda_1\eta_k}s^T(t-\eta_k)Q_1s(t-\eta_k)\\&+\eta_k^2\dot{s}^T(t)R_1\dot{s}(t)+2\lambda_1s^T(t)P_1s(t)\\&-\eta_k[s^T(t)R_1s(t)-s^T(t)R_1s(t-\eta_k)]\\&-\eta_k[-s(t-\eta_k)R_1s^T(t)+s^T(t-\eta_k)R_1s(t-\eta_k)]\\&+[\dot{s}^T(t)\times G_1+e^T(t)\times G_2]\times[Ms(t)+N\sigma(t)+Wf(t)-\dot{s}(t)]\end{aligned}\tag{22}$$

The (4) is brought into (22) and get the following.

$$\begin{aligned}\dot{V}_1(t)+2\lambda_1V(t)=&2s^T(t)P_1\dot{s}(t)+s^T(t)Q_1s(t)-e^{-2\lambda_1\eta_k}s^T(t-\eta_k)Q_1s(t-\eta_k)\\&+\eta_k^2\dot{s}^T(t)R_1\dot{s}(t)+2\lambda_1s^T(t)P_1s(t)\\&-\eta_k[s^T(t)R_1s(t)-s^T(t)R_1s(t-\eta_k)]\\&-\eta_k[-s^T(t)R_1s(t-\eta_k)+s^T(t-\eta_k)R_1s(t-\eta_k)]\\&+\dot{s}^T(t)G_1Ms(t)+\dot{s}^T(t)G_1NKe(t)+\dot{s}^T(t)[G_1+Wf(t)]\\&-\dot{s}^T(t)G_1\dot{s}(t)+e^T(t)G_1Ms(t)+e^T(t)G_1NKe(t)\\&+e^T(t)[G_1+Wf(t)]-e^T(t)G_1\dot{s}(t)\end{aligned}\tag{23}$$

According to (4), when $t\in[t_k,t_k+\eta_k)$, we can obtain $\varphi(t)<0$. If $\dot{V}_1+2\lambda_1V_1(t)\leqslant 0$ need to be proven, we can prove $\dot{V}_1+2\lambda_1V_1(t)-\varphi(t)\leqslant 0$. Thus, (23) can be optimized.

$$\begin{aligned}\dot{V}_1(t)+2\lambda_1V(t)<&2s^T(t)P_1\dot{s}(t)+s^T(t)Q_1s(t)-e^{-2\lambda_1\eta_k}s^T(t-\eta_k)Q_1s(t-\eta_k)\\&+\eta_k^2\dot{s}^T(t)R_1\dot{s}(t)+2\lambda_1s^T(t)P_1s(t)\\&-\eta_k[s^T(t)R_1s(t)-s^T(t)R_1s(t-\eta_k)]\\&-\eta_k[-s^T(t)R_1s(t-\eta_k)+s^T(t-\eta_k)R_1s(t-\eta_k)]\\&+\dot{s}^T(t)G_1Ms(t)+\dot{s}^T(t)G_1NKe(t)+\dot{s}^T(t)[G_1+Wf(t)]\\&-\dot{s}^T(t)G_1\dot{s}(t)+e^T(t)G_1Ms(t)+e^T(t)G_1NKe(t)\\&+e^T(t)[G_1+Wf(t)]-e^T(t)G_1\dot{s}(t)-e^T(t)\Omega e(t)+\alpha(t)s^T(t)\Omega s(t)\end{aligned}\tag{24}$$

According to (4), $\alpha(t)\in[\alpha_2,2\alpha_1-\alpha_2]$, we can take $\alpha(t)_{\max}=2\alpha_1-\alpha_2$.

Assuming that $\varepsilon_1=[s(t),\dot{s}(t),s(t-\eta_k),e(t)]^T$, demonstrating $\dot{V}(t)+2\lambda_1V_1(t)\leqslant\varepsilon(t)T\varepsilon^T(t)$ needs to prove $T<0$, so (24) will be written in the quadratic form as follows.

$$T=\begin{bmatrix}2\lambda_1P_1+Q_1-\eta_kR_1+2(\alpha_1-\alpha_2)\Omega & P_1+\frac{G_1M}{2} & \eta_kR_1 & \frac{G_1M}{2}\\ * & \eta_k^2R_1-G_1 & 0 & \frac{-G_1+G_1NK}{2}\\ * & * & -\eta_kR_1-e^{-2\lambda_1\eta_k}Q_1 & 0\\ * & * & * & -\Omega\end{bmatrix}\tag{25}$$

When $t\in[t_k+\eta_k,t_{k+1})$, $\sigma(t)=0$, $\varphi(t)<0$, conditions then are as follows.

$$\begin{aligned}\dot{V}_2(t)+2\lambda_1V_2(t)<&2s^T(t)P_2\dot{s}(t)+s^T(t)Q_2s(t)-e^{2\lambda_2\eta_k}s^T(t-\eta_k)Q_2s(t-\eta_k)\\&+\eta_k^2\dot{s}^T(t)R_2\dot{s}(t)-2\lambda_2s^T(t)P_2s(t)\\&-\eta_k[s^T(t)R_2s(t)-s^T(t)R_2s(t-\eta_k)]\\&-\eta_k[-s^T(t)R_2s(t-\eta_k)+s^T(t-\eta_k)R_2s(t-\eta_k)]\\&+\dot{s}^T(t)G_3Ms(t)+\dot{s}^T(t)G_3Wf(t)-\dot{s}^T(t)G_3\dot{s}(t)\end{aligned}\tag{26}$$

Assuming $\varepsilon_2=[s(t),\dot{s}(t),s(t-\eta_k)]^T$, matrix T_2 can be found in the same way.

$$T_2=\begin{bmatrix}Q_2-2\lambda_2P_2-\eta_kR_2 & P_2+\frac{G_3M}{2} & \eta_kR_2\\ * & \eta_k^2R_2-G_3 & 0\\ * & * & -\eta_kR_2-e^{2\lambda_2\eta_k}Q_2\end{bmatrix}\tag{27}$$

The following can be proven by the equation solver.

$$\begin{cases}\dot{V}_1+2\lambda_1V_1(t)\leqslant 0\\ \dot{V}_2-2\lambda_2V_2(t)\leqslant 0\end{cases}\tag{28}$$

The integral (28) can obtain the following.

$$\begin{cases}V_1(t)\leqslant e^{-2\lambda_1(t-t_k)}V_1(t_k),\ \forall t\in[t_k,t_k+\eta_k]\\ V_2(t)\leqslant e^{2\lambda_2(t-t_k-\eta_k)}V_2(t_k+\eta_k),\ \forall t\in[t_k+\eta_k,t_{k+1})\end{cases}\tag{29}$$

The following can be seen from the monotonicity.

$$\begin{aligned}&V_1(t_k)\leqslant\sigma_2V_2(t_k^-)\\&V_2(t_k+\eta_k)\leqslant\sigma_1e^{2(\lambda_1+\lambda_2)\eta_k}V_1(t_k+\eta_k^-)\end{aligned}\tag{30}$$

Assuming $S_k=t_k+\eta_k$, when $t\in[t_k,S_k]$, the following is available under the simultaneous equations of (28) (29) (30).

$$\begin{aligned}V(t)&\leqslant\sigma_2V_2(t_k^-)\leqslant\sigma_2e^{2\lambda_2(t-S_k)}\sigma_1e^{2(\lambda_1+\lambda_2)\eta_k}V_1(S_{k-1}^-)\\&\leqslant\sigma_1\sigma_2e^{2\lambda_2(t-S_k)+2(\lambda_1+\lambda_2)\eta_k}e^{-2\lambda_1(t-t_{k-1})}V_1(t_{k-1})\\&\leqslant\sigma_1\sigma_2e^{2\lambda_2(t-S_k)+2(\lambda_1+\lambda_2)\eta_k}e^{-2\lambda_1(t-t_{k-1})}\sigma_2V_2(t_{k-1}^-)\\&\quad\cdots\cdots\\&\leqslant e^{-\rho t}e^{\gamma_1(C_1+1)}V_1(0)\end{aligned}\tag{31}$$

At the same time,

$$\gamma_1=2(\lambda_1+\lambda_2)\eta_k+2\lambda_2C_3-2\lambda_1C_1+\ln(\sigma_1\sigma_2)$$

$$\begin{aligned}\rho=&-2\lambda_1[(t-t_k)+(t_{k-1}+\eta_{k-1}-t_{k-1})+\cdots+(t_0+\eta_0-t_0)]\\&+2\beta_2[(t_k-t_{k-1}-\eta_{k-1})+(t_{k-1}-t_{k-2}-\eta_{k-2})+\cdots+(t_1-t_0-\eta_0)]\\&\leqslant n(-2\lambda_1C_3+2\lambda_2C_1)\end{aligned}$$

Similarly, when $t\in[S_k,t_{k+1}]$, we can get the following.

$$V(t)\leqslant\frac{1}{\sigma_2}e^{-\rho t}e^{r_1(C_1+2)}V_1(0)\tag{32}$$

Assuming $\gamma_{\max}=\max\left\{e^{r_1(C_1+1)},\frac{1}{\sigma_2}e^{r_1(C_1+2)}\right\}$, $\varpi_1=\min\{\lambda_{\min}(P_1),\lambda\min(P_2)\}$, and $\varpi_2=\lambda_{\max}(P_1)+\eta_k\lambda_{\max}(Q_1)+\frac{\eta_k^{\ 3}}{2}\lambda_{\max}(R_1)$, the following can be obtained from the formulas (31) (32).

$$V(t)\leqslant r_{\max}e^{-\rho t}V_1(0)\tag{33}$$

Because $s^T(t)P_1s(t)\leqslant\lambda_{\min}(P_1)s(t)s^T(t)=\lambda_{\min}(P_1)\|s(t)\|^2$, the following is available.

$$V(t)\geqslant\varpi_1\|s(t)\|^2\tag{34}$$

The following can also be obtained after the scaling.

$$V_1(t)\leqslant\varpi_2\|s(t)\|^2\tag{35}$$

When $t=0$, it will be brought into equation and the following is available. (35)

$$V_1(0)\leqslant\varpi_2\|s(0)\|^2\tag{36}$$

Thus, it can be obtained by the simultaneous equations of (33) (34) (35) (36).

$$\varpi_1\|s(t)\|^2 \le V(t) \le r_{\max}e^{-\rho t}V_1(0) \le r_{\max}e^{-\rho t}\varpi_2\|s(0)\|^2 \quad (37)$$

The exponentially stable form of simplification construction can be obtained after simplifying (37).

$$\|s(t)\| \le \sqrt{\frac{\gamma_{\max}\varpi_2}{\varpi_1}}e^{-\frac{\rho t}{2}}\|s(0)\| \quad (38)$$

Therefore, it is proven that exponential stability is satisfied.

4. Numerical Simulation and Results

Concerning the two linear matrix inequalities that need to be proved in the previous design theorem, the augmented matrix K is obtained by solving the LMI toolbox in MATLAB when the conditions are met, including the values of the given coefficients α_1, α_2, α_3 are 0.3000001, 0.3, 0.1, the value of λ_1, λ_2 is 0.5, Ω is a six-dimensional identity matrix, and the value of the matrix M,N in the system error formula is given in the following formula. Finally, the augmented matrix K that meets the needs is obtained by LMI linear matrix solver. Meanwhile, whether there is a solution when the external disturbance is 0 needs to be considered.

$$M=\begin{bmatrix}0&1&0&0&0&0\\0&0&0&0&0&0\\1&0&0&0&0&0\\0&0&0&0&1&0\\0&0&0&0&0&0\\0&0&0&1&0&0\end{bmatrix}, N=\begin{bmatrix}0&0\\0.0758&0.0758\\0&0\\0&0\\0.04689&-0.04689\\0&0\end{bmatrix}$$

The following parameters related to physical pitch angle and tilt angle are shown in the following table [19, 20].

Table 2. Numeral Parameters of the 3-DOF Helicopter Platform

Parameter	Value	Parameter	Value
m_h	$1.426\ kg$	L_a	$0.66\ m$
m_t	$1.87\ kg$	L_w	$0.47\ m$
g	$9.8\ m/s^2$	J_ε	$1.0348\ kg\cdot m^2$
K_f	$0.1188\ N/V$	J_p	$0.0451\ kg\cdot m^2$
L_h	$0.178\ m$	J_t	$1.0348\ kg\cdot m^2$

Through LMI toolbox, the control gain under the given parameters can be obtained as follows.

$$K=\begin{bmatrix}-49.0434&-49.6079&-16.7503&-7.9255&-8.0167&-2.7069\\-49.0434&-49.6079&-16.7503&7.9255&8.0167&2.7069\end{bmatrix}$$

After obtaining the control gain matrix K, we list the system equation expression on MATLAB according to it and draw the curve of error value changing with time in the error system based on the system equation. At the same time, the output voltage amplitude curve and the time-dependent band diagram of each control are given to express the experimental results more intuitively.

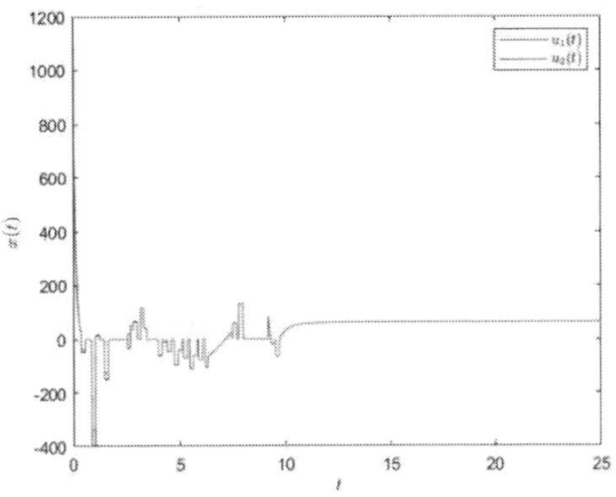

Figure 3. Output Voltage Amplitude Curve

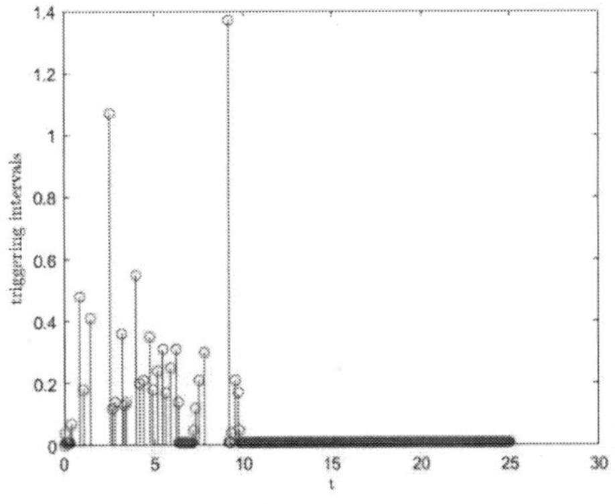

Figure 4. Triggering Intervals

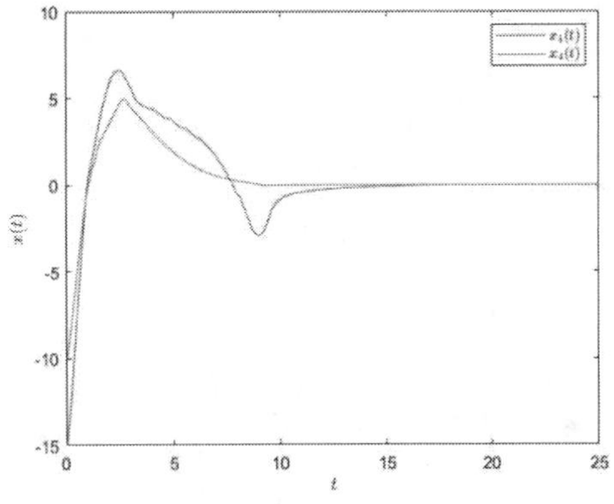

Figure 5. Systematic Error Curve

Figure 3 is a graph of output voltage amplitude, which shows that the voltage tends to be stable after the previous changes.

Figure 4 is about the time difference of triggering update, that is $t_{k+1}-t_k$, which means the time length of each update control within $25s$ in the Figure 4.1

Figure 5 is a system error curve, and the final error tends to be 0, which indicates that the system can be stabilized at last.

5. Conclusions

This paper discusses the possibility of using intermittent controller to realize the homeostatic control of 3-DOF helicopter. Besides, according to traditional methods, the stable time of the control actuator is selected by event triggering, which has good robustness, with its correctness and effectiveness verified by simulation.

References

1. Z. Li, H. H. T. Liu, B. Zhu, H. Gao, and O. Kaynak, "Nonlinear robust attitude tracking control of a table-mount experimental helicopter using output feedback," IEEE Transactions on Industrial Electronics, vol. 62, no. 9, pp. 5665–5676, 2015.
2. H. Peng and X. Zhu, Robust Attitude Control of a 3-DOF Helicopter considering Communication Delays, 2020 39th Chinese Control Conference (CCC), Shenyang, China, 2020, pp. 6780-6785, doi: 10.23919/CCC50068.2020.9188777.
3. W. Xu, H. C. Peng, Y. Liu, X. Y. Zhu, Robust attitude control of a 3-DOF helicopter prototype subject to wind disturbance and communication delay, Transactions of the Institute of Measurement and Control, vol. 43, no. 13, pp. 3071-3081
4. J. Wang, J. Wang and X. Wang, "Model Free Adaptive Sliding Mode Fault Tolerant Control for a 3-DOF Helicopter," 2022 IEEE International Conference on Unmanned Systems (ICUS), Guangzhou, China, 2022, pp. 1345-1349, doi: 10.1109/ICUS55513.2022.9986865.
5. X. Zhu, D. Li, Robust fault estimation for a 3-DOF helicopter considering actuator saturation, Mechanical Systems and Signal Processing, vol 155, no.12, p 86-102, 16 June 2021
6. L. Liu, Y. Su and J. Lu, "Attitude Tracking Control of 3-DOF Helicopter Based on Command Filter and Neural Network Techniques," 2021 IEEE 10th Data Driven Control and Learning Systems Conference (DDCLS), Suzhou, China, 2021, pp. 613-618, doi: 10.1109/DDCLS52934.2021.9455661.
7. X. D. Wang, Prescribed Performance Adaptive Fixed-Time Attitude Tracking Control of a 3-DOF Helicopter with Small Overshoot, arXiv, January 23, 2022, doi: 10.48550/arXiv.2201.09180
8. Besada-Portas, E.; Lopez-Orozco, J.A.; Aranda, J.; De La Cruz, J.M, Virtual and remote practices for learning control topics with a 3dof quadrotor, IFAC Proceedings Volumes (IFAC-PapersOnline), vol 10, PART 1, p78-83,2013,10th IFAC Symposium on Advances in Control Education, ACE 2013-Proceedings
9. Helio S. Esteban Villega, Carlos Borras Pinilla, and Laura Milena Prieto. 2021. Predictive Control of 3 DOF Helicopter Using a Kalman and Neural Network Estimator. In 2021 International Symposium on Electrical, Electronics and Information Engineering (ISEEIE 2021). Association for Computing Machinery, New York, NY, USA, 122–127. doi:10.1145/3459104.3459126
10. G. Rigatos, P. Wira, M. A. Hamida, M. Abbaszadeh and J. Pomares, "Nonlinear optimal control for the 3-DOF laboratory helicopter," 2020 IEEE 29th International Symposium on Industrial Electronics (ISIE), Delft, Netherlands, 2020, pp. 555-560, doi: 10.1109/ISIE45063.2020.9152254.
11. M. Pal, F. Plestan and A. Chriette, "Discrete-time second-order-sliding-mode observer for state and unknown input estimation: Application to a 3DOF helicopter," 2015 European Control Conference (ECC), Linz, Austria, 2015, pp. 2914-2919, doi: 10.1109/ECC.2015.7330980.
12. Alvarez-Munoz, J.; Marchand, N.; Guerrero-Castellanos, J.F.; Tellez-Guzman, J.J.; Escareno, J.; Rakotondrabe, M., Rotorcraft with a 3DOF Rigid Manipulator: Quaternion-based Modeling and Real-time Control Tolerant to Multi-body Couplings, International Journal of Automation and Computing, vol 15, no 5, p 547-58, Oct. 2018
13. J. Q. Pan, T. Shen, T. Rong, Robust Guarantee Control for 3DOF Helicopters Based on Multiple-Model Switching, Proceedings-9th International Conference on Intelligent Human-Machine Systems and Cybernetics, IHMSC 2017, vol 1, p 430-435, September 20, 2017
14. Khizer, A.N., Y. P. Dai, Ali, S.A., X. Y. Xu, 3DoF Model Helicopter with Hybrid Control, IAES TELKOMNIKA Indonesian Journal of Electrical Engineering, vol.12, no.5, p 3863-72, May 2014
15. S. M. Ra'afat, S. S. Ali and R. Akmeliawati, "Real-time optimization and control of 3DOF helicopter system via extremum seeking algorithm," 2015 10th Asian Control Conference (ASCC), Kota Kinabalu, Malaysia, 2015, pp. 1-6, doi: 10.1109/ASCC.2015.7244822.
16. R. Mei, Q. L. Cui, Backstepping control for a 3DOF model helicopter with input and output constraints, International Journal of Advanced Robotic Systems, vol.14, no.1, February 1, 2016
17. C. Yu, S. Liang, Model predictive control for a 3DOF laboratory helicopter based on disturbance prediction, ICIC Express Letters, Part B: Applications, vol.8, no.2, p 429-36, Feb. 2017
18. Ferreira, A. S. R. Caregnato-neto, A., R.K.H.; Afonso, R.J.M., Dynamic Matrix Control of a 3DOF Helicopter with Stabilising Inner Loop, CONTROLO 2022: Proceedings of the 15th APCA International Conference on Automatic Control and Soft Computing. Lecture Notes in Electrical Engineering (930), p 227-38, 20220 doi: 10.1007/978-3-031-10047-5_20

19. H. Castaneda, F. Plestan, A. Chriette, J. de Leon-Morales, Continuous differentiator based on adaptive second-order sliding-mode control for a 3-DOF helicopter, IEEE Trans. Ind. Electron. Vol.63, no.9, p 5786–5793, 2016.
20. X. Yang, X. Zheng, Adaptive NN backstepping control design for a 3-DOF helicopter: theory and experiments, IEEE Trans. Ind. Electron. Vol.67, no.5, p3967–3979, 2020.

Analysis and Discussion on Tension of Long Transport Belt

Haojie Mu

Zenith Steel Group Company Limited (Nantong), Nantong 226000 China

Abstract. The technical field of belt tensioning force studied in this paper, in particular, relates to a hydraulic belt tensioning device, including a base, a hydraulic system and a rail fixed on the base, and a tail wheel trolley sliding on the rail; This device can provide a large tension to the tail wheel of the conveyor belt to meet the tension required by the long conveyor belt. At the same time, it can automatically adjust the tension to the tail wheel of the conveyor belt according to the tension of the conveyor belt to realize the dynamic tension control of the conveyor belt; The device prevents belt deviation and material leakage, reduces labor intensity of workers, avoids equipment accidents, and improves industrial production capacity.

Key words: Long conveyor belt; Hydraulic pressure; Tension; Dynamic equilibrium.

1. Introduction

The conveyor belt was originally developed from the conveyor belt. With the continuous expansion of the use of the conveyor and the development of science and technology, the variety of the conveyor belt is expanding to meet the requirements of the conveyor. However, all conveyors may make the conveyor belt lengthen and deform after a period of operation. The lengthening of the conveyor belt is composed of elastic elongation and permanent elongation. The conveyor belt was originally developed from the conveyor belt. With the continuous expansion of the use of the conveyor and the development of science and technology, the variety of the conveyor belt is expanding to meet the requirements of the conveyor. However, all conveyors may make the conveyor belt lengthen and deform after a period of operation. The lengthening of the conveyor belt is composed of elastic elongation and permanent elongation. Therefore, a tensioning device is required to overcome the defects caused by the lengthening of the conveyor belt. Most of the tensioning devices currently used are realized by counterweights, but for long belts, the tension brought by counterweights is far from the required tensioning force [1].

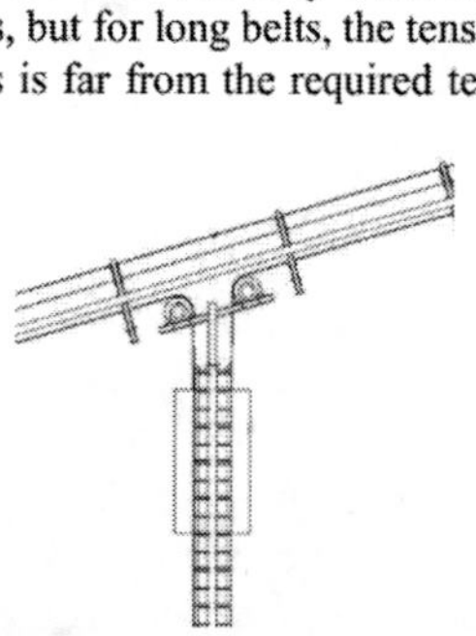

Fig. 1 Schematic diagram of conveyor belt with counterweight

2. Hydraulic tensioning scheme of conveyor belt

For a long belt, the tension brought by the counterweight is far from the required tension. In order to overcome the above shortcomings, we have conceived a hydraulic belt tensioning device that can provide a large tension to the tail pulley of the belt to meet the required tension of the long belt.

The hydraulic tensioning device of the conveyor belt comprises a base, a hydraulic system and a rail fixedly mounted on the base, and a tail wheel trolley sliding mounted on the rail. The hydraulic system comprises an oil tank, a tensioning oil cylinder and an accumulator. The tensioning oil cylinder is provided with a cylinder body, a piston rod sliding mounted in the cylinder body, and a rod cavity and a rod free cavity formed in the cylinder body. The rod cavity is connected and connected with the accumulator through the tensioning oil circuit, The rodless chamber is connected and connected with the oil tank, and the tail wheel trolley is fixedly connected with the output end of the piston rod far from the cylinder body through the connection part [2].

The hydraulic system also includes a hydraulic servo system, a main oil outlet, a pressure regulating valve, a first oil outlet, a second oil outlet, a third oil outlet and a first reversing valve. The oil inlet and outlet of the hydraulic servo system are respectively connected with the oil inlet of the oil tank and the main oil outlet. The main oil outlet is connected to the oil tank through the pressure regulating valve. The four interfaces of the first reversing valve are respectively connected with the oil outlet of the main oil outlet, the oil inlet of the first oil outlet The oil inlet of the second oil outlet circuit is connected with one port of the third oil outlet circuit, the oil outlet of the second oil outlet circuit is connected with the tensioning oil circuit, the other port of the third oil outlet circuit is connected with the rodless chamber, and one port of the third oil outlet circuit is always connected with the oil inlet of the second oil outlet circuit through the first reversing valve. When the first reversing valve is switched, the main oil circuit is connected with the second oil outlet circuit, or the main oil circuit is connected with the third oil outlet circuit.

The connecting part includes a connecting rope and a winding unit arranged on the tail wheel trolley. The hydraulic tensioning device of the conveyor belt also includes a rope winding drive mechanism fixedly installed on the base. One end of the connecting rope is fixedly installed on the rotary drive shaft of the rope winding drive mechanism, and the other end passes through the winding unit and is connected with the output end of the sliding plug rod.

The connecting part includes a connecting rope and a winding unit arranged on the tail wheel trolley. The hydraulic tensioning device of the conveyor belt also includes a rope winding drive mechanism fixedly installed on the base. One end of the connecting rope is fixedly installed on the rotary drive shaft of the rope winding drive mechanism, and the other end passes through the winding unit and is connected with the output end of the sliding plug rod. The rope winding drive module comprises a winch and a hydraulic motor for driving the rotation of the rotary drive shaft of the winch.

The hydraulic system also comprises a second reversing valve, a fourth oil circuit, a fifth oil circuit, a sixth oil circuit and a driving oil circuit. The four interfaces of the second reversing valve are respectively connected to the oil outlet of the second oil circuit, the oil inlet of the fourth oil circuit, the oil inlet of the fifth oil circuit and the oil inlet of the sixth oil circuit, The oil outlet of the fourth oil circuit is connected to the oil tank, the first oil outlet and the second oil outlet of the drive oil circuit are respectively connected with the oil outlet of the fifth oil circuit and the oil outlet of the sixth oil inlet circuit, and the drive oil circuit is provided with a hydraulic motor. When the second reversing valve is switched, the second oil outlet circuit is connected with the fifth oil outlet circuit, the sixth oil outlet circuit is connected with the fourth oil circuit, or the second oil outlet circuit is connected with the fourth oil circuit.

The rope winding drive module includes a winch and two hydraulic motors which are arranged in parallel on the drive oil circuit. When the second reversing valve is switched, the second oil circuit is connected with the fifth oil circuit and the sixth oil circuit is connected with the fourth oil circuit, or the second oil circuit is connected with the sixth oil circuit and the fifth oil circuit is connected with the fourth oil circuit, or the second oil circuit is connected with the fourth oil circuit.

The hydraulic system also includes a pressure measuring oil circuit and a pressure gauge. The pressure measuring oil circuit is connected to the tensioning oil circuit, and the pressure gauge is set on the pressure measuring oil circuit.

The hydraulic system also includes a manual ball valve, which is arranged on the pressure measuring oil circuit in turn [3].

The connecting part also includes a first pulley and a second fixed pulley which are symmetrically arranged and rotatably installed on the tail wheel trolley. The first pulley forms a winding unit. The driving end of the piston rod is rotatably installed with a movable pulley. The other end of the connecting rope bypasses the first pulley, the movable pulley and the second fixed pulley in turn and is fixed on the base.

The connecting part also comprises two rope guide pulleys which are symmetrically arranged and rotatably installed on the base, and the other end of the connecting rope bypasses one of the rope guide pulleys, the first pulley, the movable pulley and the second fixed pulley in turn and is fixed on the other rope guide pulley.

The rope winding drive mechanism also comprises a safety pin, a first safety hole is uniformly arranged on the baffle on one side of the winch, a second safety hole corresponding to the first safety hole is arranged on the base, and the safety pin is used to insert the first safety hole and the second safety hole in turn.

3. Tensioner operation theory

As shown in Figure 2, the hydraulic tensioning device of the conveyor belt includes a base, a rope winding drive mechanism, a hydraulic system and a rail fixed on the base, and a tail wheel trolley sliding on the rail.

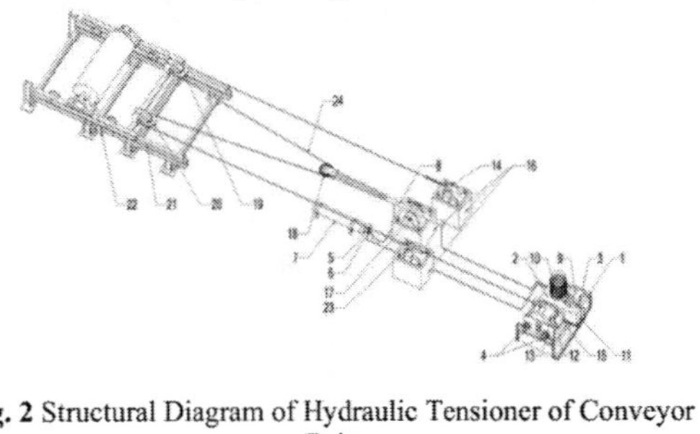

Fig. 2 Structural Diagram of Hydraulic Tensioner of Conveyor Belt

In the figure: 1. Oil tank, 2. Oil pump, 3. Second reversing valve, 4. Hydraulic motor, 5. Pressure gauge, 6. Manual ball valve, 7. Accumulator, 8. Tensioning cylinder, 9. Second reversing valve, 10. Drive motor, 11. Pressure regulating valve, 12. First safety hole, 13. Winch, 14. Second guide sheave, 15. Safety pin, 16. Base, 17. Cylinder mounting base, 18. Moving pulley, 19. Second fixed pulley, 20. First pulley, 21 Rail, 22. Tail wheel trolley, 23. First guide sheave, 24. Wire rope.

The base is a cement column fixed on the foundation.

The rope winding drive mechanism comprises a winch fixed on the base, two hydraulic motors fixed on the base and a safety pin. The hydraulic motor is connected with the winch and is used to drive the rotation of the rotary drive shaft of the winch. A plurality of first safety holes are arranged at equal intervals along the circumference of the circular plate baffle which is coaxially fixed on one side of the rotary drive shaft. A second safety hole corresponding to the first safety hole is arranged on the base. The safety pin is used to insert the first safety hole and the second safety hole in turn, so as to lock the rotary drive shaft of the winch, and prevent the connecting rope from pulling too much on the rotary drive shaft of the winch and the winch from loosening the rope.

The hydraulic system includes oil tank, tensioning cylinder, tensioning oil circuit, accumulator, hydraulic servo system, main oil outlet, pressure regulating valve, first oil outlet, second oil outlet, third oil return circuit, first reversing valve, fourth oil outlet, fifth oil outlet, sixth oil outlet, second reversing valve, second reversing valve, second reversing valve and drive oil circuit, pressure measuring oil circuit, manual ball valve and pressure gauge.

The tensioning cylinder has a cylinder block and a piston rod. The cylinder block is fixedly installed on the base through the oil cylinder mounting base. The piston end of the piston rod is slidably installed in the inner cavity of the cylinder block, so that the inner cavity of the cylinder block is divided into rod cavity and rodless cavity. The rod cavity is connected and connected with the accumulator through the tensioning oil circuit. The axial direction of the piston rod is parallel to the sliding direction of the tail wheel trolley. The axial direction of the piston rod passes through the center of the tail wheel trolley. The stroke of the tensioning cylinder is determined according to the required tension on site.

The hydraulic servo system includes oil pump and drive motor. The oil pump is immersed in the oil tank, and its oil inlet is connected with the oil tank. The output shaft of the drive motor is fixedly connected with the input end of the oil pump.

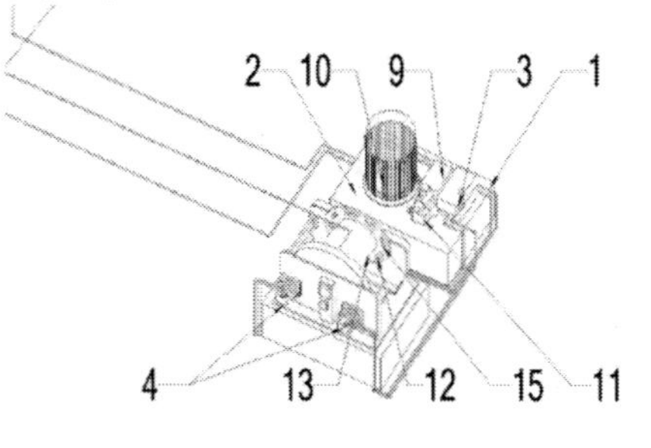

Fig. 3 Partial enlarged view of hydraulic servo system and rope winding drive mechanism

The oil inlet of the main oil outlet is connected with the oil outlet of the oil pump, and the main oil outlet is connected to the oil tank through the pressure regulating valve. The pressure regulating valve is an internal control sequence valve. The tension of the belt is controlled by adjusting the set pressure of the pressure regulating valve. When the pressure regulating valve increases the pressure, the oil pressure of the buffer cylinder will increase and the tension generated will increase.

The first reversing valve is a two position four-way electromagnetic reversing valve. The four interfaces of the electromagnetic directional valve are respectively connected with the oil outlet of the main oil outlet, the oil inlet of the first oil outlet, the oil inlet of the second oil outlet and one port of the third oil outlet.

The oil outlet of the second oil outlet is connected to the tensioning oil outlet.

The other port of the third outlet and return circuit is connected to the rodless chamber, and one port of the third outlet and return circuit is always connected with the oil inlet of the second outlet circuit through the first reversing valve.

When the first reversing valve is not energized, the main oil circuit is connected with the third oil outlet and return circuit, and the oil outlet of the second oil circuit is disconnected from the other three interfaces of the first reversing valve; When the first reversing valve is energized, the main oil circuit is connected with the oil inlet of the second oil outlet circuit.

Second directional valve Second directional valve Three position four-way manual directional valve. The four interfaces of the manual directional valve are respectively connected to the oil outlet of the second oil outlet, the oil inlet of the fourth oil return, the oil inlet of the fifth oil outlet and the oil inlet of the sixth oil inlet.

The oil outlet of the fourth return oil circuit is connected to the oil tank.

The first oil port and the second oil port of the drive oil circuit are respectively connected with the oil outlet of the fifth oil circuit and the oil outlet of the sixth oil inlet circuit, and two hydraulic motors are arranged in parallel on the drive oil circuit.

When the second reversing valve of the second reversing valve switches, the second oil outlet is connected with the fifth oil outlet and the sixth oil outlet is connected with the fourth oil outlet, or the second oil outlet is connected with the sixth oil outlet and the fifth oil outlet is connected with the fourth oil outlet, or the second oil outlet is connected with the fourth oil outlet and the fifth oil outlet and the sixth oil outlet are not connected with the second reversing valve of the second reversing valve.

The pressure measuring oil circuit is connected to the tensioning oil circuit.

The manual ball valve and pressure gauge are set on the pressure measuring oil circuit in turn. The pressure gauge is a mechanical pressure gauge. Manual ball valve is set to facilitate online replacement of mechanical pressure gauge.

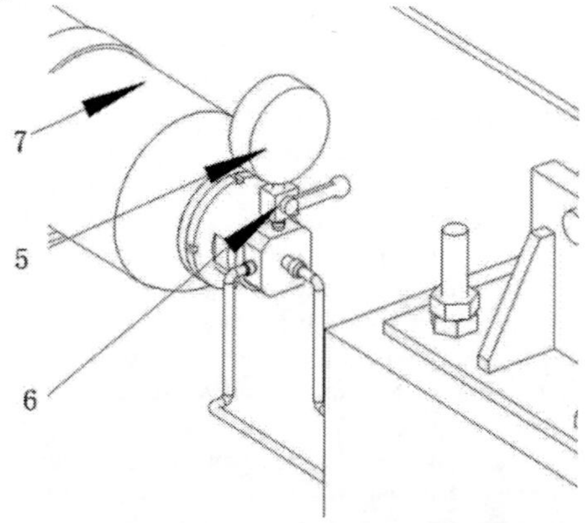

Fig. 4 Partial enlarged view at mechanical pressure gauge

The tail wheel trolley is fixedly connected with the output end of the piston rod far from the cylinder body and the rotary drive shaft of the winch through the connecting part. Specifically, the connecting part comprises a first fixed pulley, a second fixed pulley, two rope guide pulleys and a connecting rope. The driven wheel of the conveyor belt can be rotatably installed on the tail wheel trolley.

The first pulley and the second pulley are symmetrically arranged relative to the central axis of the piston rod and can be rotatably installed on both sides of the tail wheel trolley.

The two rope sheaves are symmetrically arranged relative to the central axis of the piston rod and are rotatably installed on the base. The two rope guide pulleys are the first rope guide pulley and the second rope guide pulley. The first rope guide pulley and the first pulley are located at one side of the piston rod; The second rope guide pulley and the second fixed pulley are located on the other side of the piston rod.

The connecting rope is a steel wire rope, one end of which is fixed on the rotary drive shaft of the rope winding drive mechanism, and the other end is fixed on the second rope guide pulley after bypassing the first rope guide pulley, the first pulley, the movable pulley and the second fixed pulley. The steel wire rope is supported by the first guide sheave to prevent the steel wire rope from falling to the ground due to looseness. The first pulley and the second pulley are respectively located on both sides of the tail wheel trolley to facilitate the connecting rope to pull the tail wheel trolley.

The operating principle of the hydraulic tensioning device for the conveyor belt is: when the electromagnetic reversing valve is not energized, start the oil pump, connect the second oil outlet with the fifth oil outlet and the sixth oil outlet with the fourth oil outlet through the manual reversing valve, drive one of the hydraulic motors to rotate forward, and the hydraulic horse to drive the winch to rotate forward, tighten the wire rope, pull the tail wheel trolley backward to a proper position, so as to tighten the conveyor belt, The second oil outlet can also be connected with the sixth oil outlet and the fifth oil outlet can be connected with the fourth oil outlet through the manual directional valve, which can drive the motor to reverse and adjust the position of the tail wheel trolley [4].

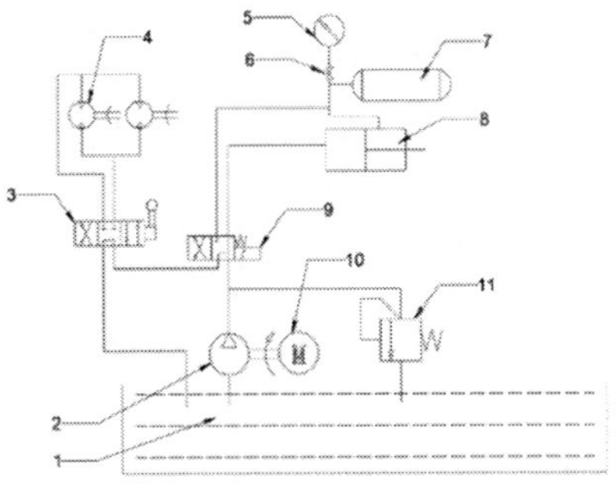

Fig. 5 Structural schematic diagram of hydraulic system

When the electromagnetic reversing valve is powered on, the main oil circuit will supply oil to the rod cavity of the accumulator and tension cylinder. The pressure regulating valve controls the oil supply pressure. Manual ball valves and mechanical pressure gauges are installed at the accumulator to feed back the oil supply pressure; When the rod cavity of the tensioning cylinder is supplied with the pressure set by the pressure regulating valve, the control oil pump stops and the electromagnetic reversing valve is powered off, the rod cavity of the accumulator and the tensioning cylinder will be closed, and the internal oil pressure is equivalent to the pull force, because the tension rod of the tensioning cylinder is connected to the tail wheel trolley, and the belt tension is constantly changing. When the belt tension becomes larger, the oil with the rod cavity will enter the accumulator, and when the belt tension becomes smaller, the oil in the accumulator will enter the rod cavity to form dynamic tension [5].

4. Conclusion

To sum up, the hydraulic tensioning device of the conveyor belt can provide a large tension to the tail pulley of the conveyor belt to meet the tension required by the long conveyor belt. The tension to the tail pulley of the conveyor belt can be automatically adjusted according to the tension of the conveyor belt to achieve dynamic tensioning control of the conveyor belt. The device prevents belt deviation and material leakage, reduces labor intensity of workers, avoids equipment accidents, and improves industrial production capacity.

References

1. Jiang Weiliang, Han Dongjin. Current situation and development trend of coal mine belt conveyor in China [J] Shanghai: Coal Mine Electromechanical, 2008:85-86.
2. Li Kun. Development of typical belt conveyor design system [D]. Beijing: North China Electric Power University, 2006: 75-76.
3. Zhao Yuwen, Li Yunhai. Current Situation and Development Trend of Belt Conveyor [J]. Beijing: Coal Mining Machinery, 2004: 158-163.
4. Wang Xiankui. Mechanical Manufacturing Technology [M] Beijing: China Machine Press, 2006:35-36.
5. Shan Ziming. Current situation and gap of belt conveyor technology [J]. Nanjing: Market Weekly, 2009: 12-14.

Feasibility study of longitudinal transportation

Zhejun Shu, Haojie Mu

Zenith Steel Group Company Limited (Nantong), Nantong 226000, China

Abstract. The research in this paper relates to the technical field of light load transportation, in particular to an electric upstairs car, which includes a frame, the top of the frame is fixedly equipped with a handle, the bottom of the frame is rotationally equipped with a group of wheels, the side wall near the bottom of the frame is fixedly equipped with a carrier plate, the support plate is fixedly arranged between the frames, and the box is fixedly arranged on the side of the support frame away from the carrier plate, A support rod is rotated on the opposite side of the box. The length of the support rod is larger than the diameter of the wheel. An arc surface is set on the support rod. A drive component is set in the box. The drive component is used to drive the support rod to rotate. This kind of electric car that can be transported longitudinally can reduce labor.

Key words: Direction; Transport; Rotating support; Electric trolley.

1. Introduction

In industrial production, goods moved in a short distance and a small range are transported by trolleys.

The well-known double-sided trolley includes a frame. The bottom of the frame is rotated and set with a set of wheels, the top of the frame is fixed with a handle, and the side wall near the bottom of the frame is set with a bottom movable baffle. Place the goods on the bottom interactive baffle. The bottom movable baffle supports the goods and pushes the handle. The wheels roll on the ground to drive the frame to move. The frame movement drives the bottom movable baffle to move, and then the movable baffle drives the goods to move. [1]

Many times, goods need to be moved from low to high or from high to low, but many industrial buildings are not equipped with elevators and only stairs. Because the wheels are not easy to go up and down the stairs, so it is necessary to carry goods manually, resulting in an increase in labor.

Fig. 1 Schematic diagram of double-sided trolley

2. Technical proposal for electric upstairs trolley

This electric upstairs trolley includes a frame. The top of the frame is fixed with a handle. The bottom of the frame is rotated with a set of wheels. The side wall near the bottom of the frame is fixed with a load plate. The support plate is fixed between the frames. The side of the support frame away from the load plate is fixed with a box. The opposite side of the box is rotated with a support rod. The length of the support rod is greater than the diameter of the wheel, The support rod is provided with an arc surface, and the box body is provided with a drive component, which is used to drive the support rod to rotate. [2]

By adopting the above technical scheme, the goods are placed on the loading plate and tied with the frame by ropes. When the car moves on the flat ground, the support rod has an arc surface on the surface, which makes the center of gravity of the support rod change in an inclined state. The end wall of the support rod is not easy to contact with the ground, thus facilitating the staff to push the car on the ground. When the car needs to go up the stairs, start the drive assembly, The drive assembly drives the support rod to rotate forward. The rotation of the support rod makes one end of the support rod contact with the upper step first, and then makes the support rod drive the frame to move upward. When the car needs to go down the stairs, start the drive assembly, and the drive assembly drives the support rod to reverse. One end of the support rod contacts with the lower step, and then the support rod drives the frame to move downward, so as to achieve the effect that the car is convenient for going up and down the stairs, and then facilitate the goods handling, Reduce the labor intensity of the staff.

The drive assembly includes a drive motor, a drive gear and a long pin shaft. The drive gear rotates and is arranged in the box. The long pin shaft penetrates the drive gear and is fixedly connected with the drive gear. The long pin shaft penetrates the box and is rotationally connected with the box. One end of the long pin shaft is fixedly connected with one of the supporting rods, the other end of the long pin shaft is fixedly connected with the other supporting rod, and the drive motor is fixedly arranged outside the box, The drive motor is used to drive the drive gear to rotate.

Start the drive motor, the drive motor drives the drive gear to rotate, the drive gear rotates to drive the long pin shaft to rotate, and the long pin shaft rotates to drive the support rod to rotate synchronously, so as to achieve the effect that the drive component drives the support rod to rotate.

The drive assembly also includes a drive gear and a reduction gear. The drive shaft of the drive motor runs through the side wall of the box and is rotationally connected with the drive gear. The reduction gear is rotationally arranged in the box. The end wall of the reduction gear is rotationally connected with the inner wall of the box. The drive gear and the reduction gear mesh with each other. The reduction gear and the transmission gear mesh with each other.

Start the drive motor, the drive motor drives the drive gear to rotate, the drive gear rotates to drive the reduction gear to rotate, and the reduction gear rotates to move the drive gear to achieve the effect of driving the drive motor to drive the drive gear to rotate.

Because the size of the drive gear is larger than the size of the reduction gear and the size of the reduction gear is larger than the size of the drive gear, when the size of the drive gear, the reduction gear and the linear speed of the drive gear are the same, the angular speed of the drive gear is greater than the angular speed of the drive gear, and then the rotational speed of the drive motor is greater than the rotational speed of the support rod. By setting the reduction gear, the rotational speed of the support rod is reduced, Reduce the difficulty of staff to control the trolley due to the fast rotation of the support rod.

The handle is provided with a start/stop switch, a reverse switch and a forward switch. The start/stop switch is used to control the start/stop of the drive motor, the forward switch is used to control the forward rotation of the motor, and the reverse switch is used to control the reverse rotation of the motor.

When the staff pushes the trolley up the stairs, the staff presses the start-stop switch and the forward switch to drive the motor forward to drive the support rod forward. When the staff pushes the trolley down the stairs, the staff presses the start-stop switch and the reverse switch to drive the motor reverse to drive the support rod reverse.

The handle is fixedly provided with an anti-skid layer. The anti-skid layer increases the friction between the handle and the hands of the staff to facilitate the staff to push the trolley. The end walls of the support rods are fixedly provided with rubber layers. The rubber layer on the end wall of the support rod increases the friction between the end wall of the support rod and the step, which is convenient for the support rod to drive the trolley up and down the stairs.

Several reinforcing plates are fixedly arranged between the frames. The reinforcing plate increases the support force of the frame to the goods, reduces the breaking of the frame during the process of goods going up the stairs, reduces the breaking of the frame during the process of goods going down the stairs, and improves the strength of the frame.

3. Electric trolley operation theory and implementation plan

The implementation principle of this electric upstairs car is: when the car moves on the flat ground, the staff pushes the handle, the handle drives the wheel to roll, and then the wheel drives the frame, load plate and goods to move. When the car needs to go up the stairs, the staff presses the start-stop switch and the forward switch to drive the motor to drive the support rod forward. The rotation of the support rod makes one end of the support rod contact with the upper step first, and then the support rod drives the frame to move up. When the car needs to go down the stairs, the staff presses the start/stop switch and the reverse switch, and the drive motor drives the support rod to reverse. One end of the support rod contacts the lower steps, and then the support rod drives the frame to move downward.

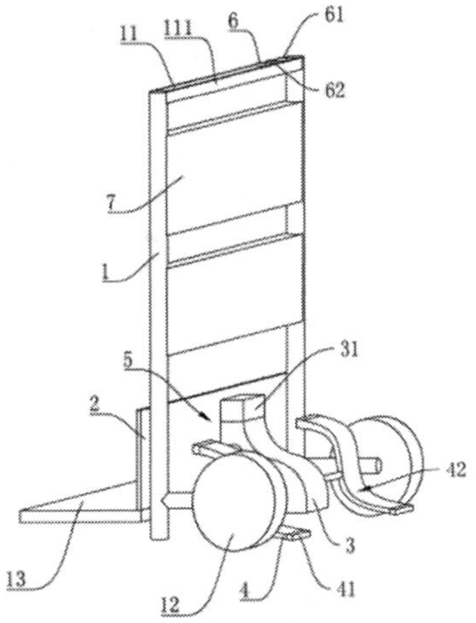

Fig. 2 Structure diagram of electric upstairs trolley

In the figure: 1. Frame; 11. Handle; 111. Anti-slip layer; 12. Wheel; 13. Load plate; 2. Support plate; 3. Box; 31. Control box; 4. Support rod; 41. Arc surface; 42. Rubber layer; 5. Drive components; 51. Drive motor; 52. Drive gear; 53. Reduction gear; 54. Transmission gear; 55. Long pin shaft; 6. Start and stop switch; 61. Reverse switch; 62. Forward switch; 7. Reinforcing plate.

This electric car for going upstairs. Referring to Figure 2, it includes a frame, which is welded by round steel pipe. A handle is installed on the top of the frame, and a group of wheels are set symmetrically at the bottom of the frame. A load plate is installed horizontally on the side wall near the bottom of the frame, and one end of the load plate is welded with the frame. The support plate is installed on the load plate along the height direction. The bottom end of the support plate is welded with the top wall of the load plate, and the side wall of the support plate is welded with the frame. The box is installed on the side of the support frame away from the load plate, and the support rod is rotated on the back of the box. The length of the support rod is greater than the diameter of the wheel. The supporting rods are symmetrically arranged, and arc surfaces are set on the supporting rods. The supporting rods are S-shaped. The box body is equipped with a drive component, which is used to drive the support rod to rotate. Place the goods on the loading plate and bind the goods with the frame by ropes. When the car moves on the flat ground, the staff pushes the handle, which drives the wheel to roll, and then the wheel drives the frame, load plate and goods to move. Since the surface of the support rod is provided with an arc surface, the center of gravity of the support rod is changed to be inclined, and the end wall of the support rod is not easy to contact the ground when the trolley moves on the flat ground, which is convenient for the staff to push the trolley on the ground. When the car needs to go up the stairs, start the drive assembly. The drive assembly drives the support rod to rotate forward. The rotation of the support rod makes one end of the support rod contact with the upper step first, and then the support rod drives the frame to move upward. When the car needs to go down the stairs, start the drive assembly. The drive assembly drives the support rod to reverse. One end of the support rod contacts the lower step, and then the support rod drives the frame to move downward. To achieve the effect that the car is easy to go up and down the stairs, thus facilitating the goods handling and reducing the labor intensity of the staff.

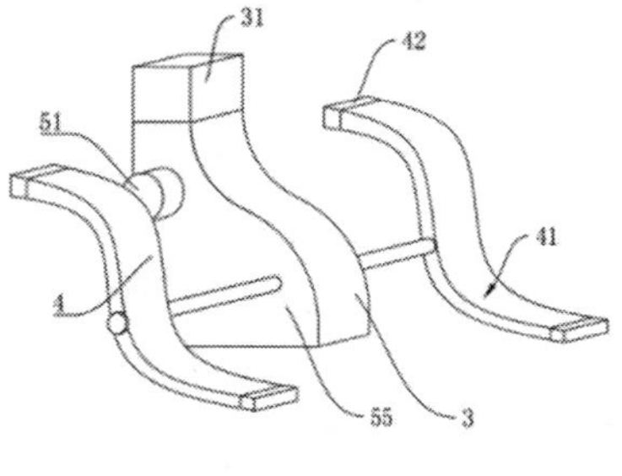

Fig. 3 Schematic diagram of connection between support plate and box

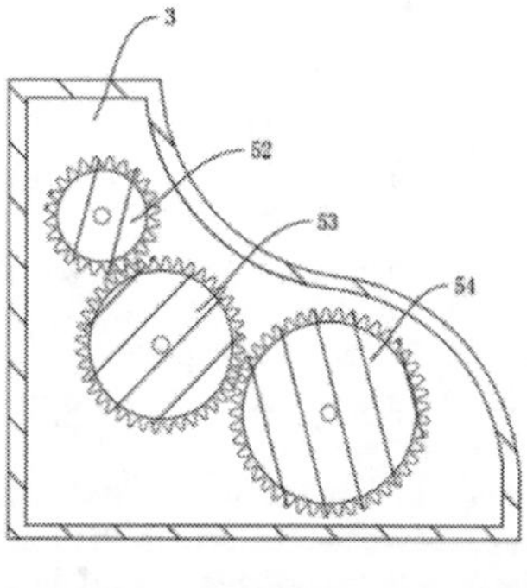

Fig. 4 Sectional view of drive assembly structure

In order to drive the support rod to rotate, refer to Figures 2 to 4. The drive assembly includes a drive motor, a drive gear, a reduction gear, a drive gear and a long pin shaft. The drive motor is installed on one side of the box, and the drive shaft of the drive shaft runs through the side wall of the box. One end of the drive gear is welded with the drive shaft of the drive motor, and the other end of the drive gear is rotationally connected with the inner wall of the box through the bearing. Both ends of the reduction gear are installed with bearings, which are rotationally connected with the inner wall of the box. The transmission gear is set in the box through the rotation of the long pin shaft. The long pin shaft runs through the transmission gear and is welded with the transmission gear. The long pin shaft runs through the box and is connected with the box through the rotation of the bearing. One end of the long pin shaft is connected with the central thread of one of the support rods and is fixed by nuts. The other end of the long pin shaft is connected with the central thread of the other support rod and is fixed by nuts. The reduction gear and the drive gear mesh with each other, and the reduction motor and the drive gear mesh with each other. Start the drive motor. The rotation of the drive motor drives the rotation of the drive gear. The rotation of the drive gear drives the rotation of the reduction gear. The rotation of the reduction gear drives the rotation of the drive gear. The rotation of the drive gear drives the rotation of the long pin shaft. The rotation of the long pin shaft drives the rotation of the support rod.

Referring to Figure 4, the size of the transmission gear is larger than that of the reduction gear, and the size of the reduction gear is larger than that of the drive gear. Because the size of the drive gear is larger than the size of the reduction gear and the size of the reduction gear is larger than the size of the drive gear, when the size of the drive gear, the reduction gear and the linear speed of the drive gear are the same, the angular speed of the drive gear is greater than the angular speed of the drive gear, and the rotational speed of the drive motor is greater than the rotational speed of the support rod. By setting the reduction gear, the rotation speed of the support rod is reduced, and the situation that the staff can not easily control the trolley due to the fast rotation of the support rod is reduced.

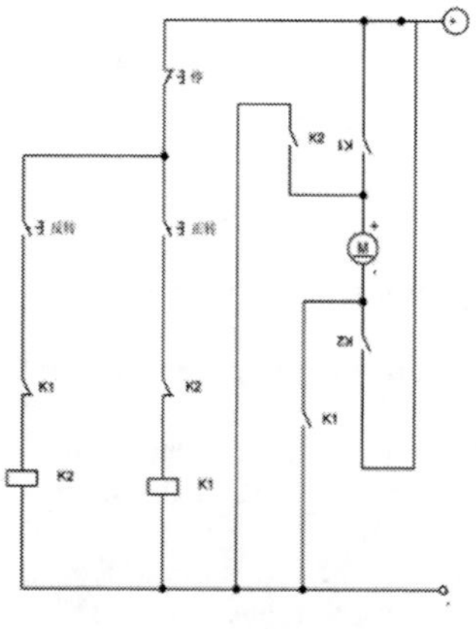

Fig. 5 Circuit control diagram of positive and reverse rotation of motor

In order to facilitate the staff to control the rotation of the support rod, refer to Figure 2 and Figure 5. A control box is installed on the box. The battery, forward relay K1 and reverse relay K2 are installed in the control box. The positive end of the battery is connected with the positive connector of the drive motor, and the negative end of the battery is connected with the negative connector of the drive motor. The forward relay K1 and the reverse relay K2 are interlocked. The handle is equipped with a start-stop switch, a forward switch and a reverse switch. The start-stop switch, a forward switch and a reverse switch are connected in the circuit through wires, and the wires are routed through the round steel pipe of the frame. The start-stop switch is used to control the on-off of the drive motor and the battery, the forward switch is used to control the on-off of the forward relay K1, and the reverse switch is used to control the on-off of the reverse relay K2. When the staff pushes the trolley up the stairs, the staff presses the start/stop switch and the forward switch, the forward relay K1 is connected and the reverse relay K2 is disconnected, thus driving the motor forward to drive the support rod forward. When the staff pushes the trolley down the stairs, the staff presses the start-stop switch and the reverse switch, the reverse relay K2 is connected and the forward relay K1 is disconnected, which drives the motor to reverse and drives the support rod to reverse. [4] In order to facilitate the staff to push the trolley up and down the stairs, refer to Figure 1. The handle surface is pasted with anti-slip layer, which is made of rubber material. The anti-skid layer increases the friction between the handle and the hands of the staff to facilitate the staff to push the trolley. The end wall of the support rod is pasted with a rubber layer. The rubber layer on the end wall of the support rod increases the friction between the end wall of the support rod and the step, which is convenient for the support rod to drive the trolley up and down the stairs. A group of stiffening plates are installed between the frames, and the end wall of the stiffening plate is welded with the opposite side of the frame. The reinforcing plate increases the support force of the frame to the goods, reduces the breaking of the frame during the process of goods going up the stairs, reduces the breaking of the frame during the process of goods going down the stairs, and improves the strength of the frame.

4. Conclusion

To sum up, this kind of electrically driven car that can be transported longitudinally has a simple structure. By setting the drive motor, drive gear, drive gear, reduction gear and long pin shaft, the drive component drives the support rod to rotate, which is convenient for going up and down the stairs, thus facilitating the goods handling and reducing the labor intensity of the staff.

References

1. Yang Xiaolun. Development trend and prospect of electric vehicle technology [J]. Wuhan: Automotive Science and Technology. 2007 (06).
2. Yang Xiaolun, Liu Xiaokang, Wang Bin. Development trend and prospect of electric vehicle technology [J]. Beijing: Inverter World. 2007 (07).
3. Shao Jie, Zhang Yong. Discussion on the application of automation technology in the use of construction machinery [J]. Beijing: China Petroleum and Chemical Standards and Quality, 2011 (9).
4. Wang Xiankui. Machine Manufacturing Technology [M] Beijing: Machine Industry Press, 2006:35-36.

Multifunctional properties and applications of ultra-light porous metal materials

Jianru Chen, Da Zhang

School of Materials Science and Engineering, Shandong University of Technology, Zibo City, Shandong Province, 255000, China

Abstract. Porous metallic material is a new type of material with a special structure, which undergoes structural changes under certain conditions, thus forming a new structural unit. Due to their unique porosity and pore size structure, ultralight porous metallic materials have unique multifunctional properties, such as absorption and buffering of vibration energy, energy absorption and weight reduction, thermal insulation, vibration and noise reduction, and tunable performance, and thus are widely used in aerospace, energy industry, transportation, and biomedical fields. This paper reviews the research progress of ultra-light porous metal materials at home and abroad in recent years, including the basic structure and main properties of ultra-light porous metal materials, and discusses their application fields. Although China started late in this field, it has taken a place in the international arena with the continuous research work. In the future, we should focus on basic research and engineering application research to lay a solid foundation for China to achieve technological breakthrough in this field.

Keywords: ultra-light porous metal materials; functional properties; application analysis.

Ultra-light porous metal material is a kind of metal material with special structure, which has a large number of internal pores, most of which are connected, while a small portion of pores are connected or partially connected. As the porous metallic material will undergo structural changes under certain conditions, thus forming a new structural unit with unique multifunctional properties, such as absorbing and buffering vibration energy, energy absorption and weight reduction, heat insulation, vibration and noise reduction, and adjustable performance. Therefore, the material has broad application prospects in the fields of aerospace, transportation, energy industry and biomedicine. At the same time, due to its unique structural characteristics and performance advantages, it is widely used in manufacturing equipment and special equipment, such as the aerospace field for the manufacture of flight control systems, thermal management systems and anti-vibration systems; transportation field for the manufacture of aircraft engine blades, aircraft engine fans and turbocharger blades; energy industry for the manufacture of gas turbines and nuclear reactor fuel cladding and combustion chamber The biomedical field is used to manufacture artificial heart valves and medical implantable artificial organs.

1. Basic structure and performance

The basic structural unit of ultra-light porous metal materials is a porous structure obtained by a certain composition of metal monomers, in accordance with certain process requirements, by melting and casting, rolling and other processes. It mainly includes composite structural materials made of metal matrix and porous materials and direct composite structural materials made of metal matrix and porous materials.

The main properties of ultralight porous metallic materials are closely related to their porosity, pore size and distribution, phase interface and surface state. Due to their unique physical, mechanical and thermodynamic properties, they are widely used in aerospace, energy industry, transportation and biomedical fields. Ultra-light porous metallic materials can be divided into small pore size metallic materials and large pore size metallic materials according to the pore size. Small pore size metallic materials are generally made of high porosity matrix (such as aluminum and aluminum alloy) and low porosity porous phase (such as austenitic stainless steel) composite, its structure and properties are shown in Table 1. large pore size metallic materials are made of low porosity matrix (such as carbon fiber, glass fiber, aramid fiber, etc.) and high porosity porous phase (such as pure aluminum, magnesium alloy) composite.

1.1 Mechanical properties

Because of their unique pore structure, porous metallic materials can achieve properties with different strengths, different deformation patterns, and different forms of energy absorption. Compared with traditional metallic materials, porous metallic materials have significant advantages in terms of strength and stiffness. The fracture toughness of ultralight porous metallic materials is much higher than that of traditional metallic materials, and they also have good damping properties and energy absorption characteristics, so they can effectively absorb energy, absorb vibration energy, and do not cause damage to the surrounding environment. In addition, it also has good dimensional stability and good processability.

At present, ultra-light porous metal materials are mainly used in the aerospace, energy industry and transportation fields. Among them, the aerospace field is mainly used in aircraft fuselage and other structural components, spacecraft hull and engine combustion chamber and other parts; the energy industry is mainly used in wind power blades and battery shells.

1.2 Energy absorption and weight reduction

In the process of car collision, due to the impact load, the body will be deformed, if the energy absorption capacity is poor, it will lead to serious deformation of the body or even a car damage accident. In order to solve this problem, a large number of energy-absorbing and weight-reducing devices need to be installed on the body. If the traditional structural design method is used, a large number of additional parts and materials are needed to achieve energy absorption and weight reduction. However, the use of ultra-light porous metal materials for energy absorption and weight reduction can achieve structural lightweighting because of its excellent energy absorption and weight reduction performance.

Since the pore structure of ultralight porous metal materials is irregular, the lap between different pore structures of ultralight porous metal materials can be used to achieve the energy absorption and weight reduction function. In this way, energy absorption, volume and weight reduction as well as cost reduction can be achieved, so it has great application prospects. At present, the research on energy absorption and weight reduction of porous metal materials mainly focuses on reducing the density of porous metal materials to achieve. By optimizing the design of porous metal materials, the density can be reduced to about 1/5 of the original, so that the effect of energy absorption and weight reduction can be effectively improved.

Research shows that when the car collision speed reaches 50 km/h, the maximum deformation of the body is 0.075 mm without any additional parts and materials, while when the collision speed reaches 100 km/h, the maximum deformation of the body is 0.001 mm, which indicates that the energy absorption capacity of the body is greatly improved.

1.3 Insulation

Due to their high thermal conductivity, ultralight porous metallic materials are widely used in aerospace applications. According to the International Council for Thermal Transport, porous metallic materials are defined as porous structures with a pore size of less than 20 μm, while metallic materials are solids composed of pure metals or alloys. In the aerospace field, metallic materials can absorb and consume most of the heat and provide thermal insulation.

For ultra-light porous metal materials, their thermal conductivity is much higher than that of ordinary metal materials of the same density, but their thermal conductivity is only 1/3~1/2 of that of carbon fiber of the same mass, which has a great advantage in thermal insulation. Usually, when designing porous metallic materials using traditional thermal insulation methods, the thermal conductivity inside the material is difficult to determine. When porous structure design is used, there will be a large free energy potential difference at the atomic-molecular level due to the presence of a large number of free electrons and ions in the pore space, which makes the material obtained by porous structure design have a high thermal conductivity. Therefore, how to improve the thermal insulation performance of porous metallic materials is one of the hot research topics in this field.

At present, many researchers have introduced new thermal insulation materials into the design of porous metal materials, the most representative of which is porous carbon fiber composite. This material has extremely high thermal insulation performance, but the relatively heavy carbon fiber makes the comprehensive cost of the material high, so few people use this material for porous metal material design. Combining ultralight porous metal materials with carbon fiber composites can effectively reduce the cost of the material and improve its comprehensive performance. As shown in Figure 6, the combination of ultralight porous metal and carbon fiber composite material can not only reduce the cost of the material, but also improve the comprehensive performance of the material, which meets the requirement of comprehensive performance of metal insulation material in aerospace field.

1.4 Vibration and noise reduction

Porous metal materials have a unique porous structure, and their porosity and pore size structure can effectively absorb vibration energy, and thus have vibration and noise reduction properties. Due to their unique pore structure, porous metal materials can be used in many fields. For example, they can be used in the aerospace industry for vibration and noise reduction, and in automobiles and ships for vibration isolation and damping.

Due to the unique aperture structure of ultra-light porous metal materials, they have great potential in vibration energy absorption. When using porous metallic materials as sound insulation materials, not only can structural acoustic radiation noise be reduced, but also the sound insulation effect of the sound insulation layer can be improved. For example, the combination of ultralight porous metallic materials with conventional materials can significantly reduce structural acoustic radiation noise and structural vibration noise. When using ultra-light porous metal materials as sound insulation materials, the sound source can be effectively isolated from the sound insulation layer, thus achieving the effect of noise reduction. At the same time, porous metal materials also have good performance advantages in vibration absorption. When porous metal materials are used as vibration absorbing materials, they can effectively absorb structural vibration energy and reduce structural vibration noise. When the porous metal material has a porous structure, its vibration absorption capacity is obviously improved.

Therefore, porous metal materials have great potential for application in vibration and noise reduction. In addition, with the rapid development of new high-tech manufacturing industry and the increasing attention to noise pollution control in various countries, the vibration and noise reduction performance of ultra-light porous metal materials has put forward higher requirements. Therefore, the study of new high-tech manufacturing industry with ultra-light porous metal materials is of great significance to solve the vibration and noise problems faced by the aerospace, transportation and energy industries.

2. Second, the application analysis of ultra-light porous metal materials

2.1 Application to the transportation and manufacture of mechanical parts

Porous metallic materials are characterized by high strength, low density, and light weight, and are naturally superior in the delivery and manufacture of mechanical components. For example, an extremely light, porous metallic material can be used as a raw material for steel plates, and the steel plates produced from this raw material have the same light weight, durability, and performance as conventional metallic materials. In addition, the present invention allows the application of this extremely light, porous metal material to mechanical skins and automotive parts, thereby achieving a reduction in their mass.

2.2 isolation noise application

It has a large number of holes, which are filled with air. Therefore, the ultra-light porous metal material has a better sound insulation effect and can be used as a sound insulation material, which can be used in noisier roofs and can effectively stop the transmission of noise. In the near future, porous metal materials can also be used on vehicles so that engine noise can be reduced to a minimum, which has positive practical implications for chain maintenance.

2.3 Communication business

Due to its special energy-absorbing properties, it has been widely used in communication engineering. It has better wave-absorbing performance than other conventional materials, which can improve the communication quality and enhance the communication environment, making communication more convenient and smooth.

2.4 Environmental Engineering

The porous metal material is also sensitive to some gases, so it can play a role in environmental engineering. This method can be applied in areas such as environmental monitoring. From the reaction of a porous metal with a certain gas, it is possible to tell whether there is environmental pollution in this area, and also to determine the source of the pollution from the detected gas.

2.5 Bone material

In addition to the above characteristics, porous metallic materials have properties and structures similar to those of biological materials; therefore, they do not cause rejection when used as bone repair materials, and therefore, they have broad clinical application prospects. In future research, porous metallic materials will become an ideal bone material.

2.6 as a plate

Extremely light porous metallic materials are also used in military applications. The ultra-lightweight metal material takes advantage of the material's strength, stiffness and stretchability to create a stainless steel sandwich structure. This material is lightweight, stiff and durable. Metal panels made from this material have properties such as heat insulation, heat dissipation, heat resistance, and sound absorption. Nowadays, military fire partitions are made of perforated, extremely light metal sheets to avoid heat and sound disturbance to the occupants in the cabin.

2.7 for anti-electromagnetic interference

Power electronic devices and systems, on the one hand, are very sensitive to electromagnetic interference in the environment; at the same time, they also generate a certain amount of electromagnetic interference to the surrounding environment. Therefore, electromagnetic shielding has also become a hot topic At present, China has made some initial progress in using lightweight porous materials to absorb waves, but its specific applications are yet to be studied and analyzed in depth.

2.8 Other areas

In the aerospace field, ultra-light porous metal materials can absorb vibration energy, reduce structural mass, increase structural stiffness and improve starting performance. In the energy industry, ultra-light porous metal materials can be used as electromagnetic shielding, heat shielding materials, absorb vibration energy and radiant heat; in the transportation field, ultra-light porous metal materials can be used for vibration isolation, cushioning and energy absorption inside the car; in the biomedical field, ultra-light porous metal materials can be used to reduce sound pressure, absorb and release sound energy; in the aerospace field, ultra-light porous metal materials can be used for noise reduction and heat insulation. In the field of aerospace, ultra-light porous metal materials can be used for noise reduction and heat insulation. Many units have already conducted research on ultra-light porous metal materials.

3. Concluding remarks

To conclude, porous metal material is a new material obtained through long time research, which has more features and functions than traditional materials, and can combine the double advantages of structural and functional materials. With the continuous development of society, our living space will become smaller and smaller, and resources will become less and less. The only way to ensure that our life will not be threatened is to find more efficient and environmentally friendly materials through in-depth research on new materials. In addition, with the development of science and technology, we already have the ability to conduct in-depth research and development on porous metal materials, which will become an inevitable trend for human survival in the future. Only through continuous innovation and discovery of various uses and applications of porous metal materials can we bring more convenience to our life and more vitality to our life.

References

1. T.J. Lu, D.P. He, C.Q. Chen, et al. Multifunctional properties and applications of ultralight porous metallic materials [J]. Advances in Mechanics, 2006, 36(4): 517-535.
2. Hou Xiuhui. Optimal design of porous metallic materials and its new characterization of wave propagation problem [D]. Xi'an: Northwestern Polytechnical University, 2010.
3. Yan Jun. Multi-scale analysis and collaborative optimization design of ultralight metal structures and material properties [D]. Dalian: Dalian University of Technology, 2007.
4. Qiu Liang, Liu Yaohuan. Ultralight porous metal materials for military applications[J]. Science and technology innovation and application, 2014, (22): 83-83.

MATEC Web of Conferences **380**, 01027 (2023) https://doi.org/10.1051/matecconf/202338001027
MSMS 2023

Comparisons of Design Schemes for the upper beam support at east cable tower of Lingdingyang Bridge on Shenzhen-Zhongshan Link

Suhai Yang [1,2,*], Jun Zhao [1,2], Zhen Li [1,2], Lulu Luo [1,2]

[1] CCCC Second Harbor Engineering CO., LTD., Wuhan 430040, China
[2] Key Laboratory of Large-span Bridge Construction Technology, Ministry of Communications, PRC., Wuhan 430040, China

Abstract. The complex site conditions of Lingdingyang Bridge on Shenzhen-Zhongshan Link lead to many construction difficulties. The design of the support affects the construction quality of the super-height, large-weight and long-span beam on the cable tower. In order to achieve a high-standard construction and control the quality of the project, three support design schemes for the upper beam of the east cable tower of Lingdingyang Bridge were studied and compared by calculating and analyzing with the finite element software. Ultimately, the plan of the upper arch bracket stands out among the three schemes. The bracket scheme can not only meet the quality requirement of the main structure, but also reduce the construction difficulty and cost of the temporary structure, which can provide reference for the design of similar projects in the future.

Keywords: Shenzhen-Zhongshan Link, cable tower, upper beam support, comparisons.

1. Introduction

Shenzhen-Zhongshan Link is the first cluster project in the world integrating ultra-wide and ultra-long submarine tunnels, super-large cross-sea bridges, deep-water artificial islands and underwater interchange, with unprecedented scale and technical difficulty [1]. Shenzhen-Zhongshan Link (Shenzhong Link) connects Jihe Expressway in the east, crosses the Pearl River Estuary, and reaches Zhongshan Maanshan Island in the west. It is joined to the planned Zhongkai Expressway and the Eastern Outer Ring Expressway, and lands in Shenzhen, Zhongshan and Nansha in Guangzhou. Shenzhong Link is about 24km long, among which the sea crossing section is 22.39km long. The project adopts the technical standard of eight-lane expressway in two ways with a design speed of 100km/h.

The key route of Shenzhong Link is Lingdingyang Bridge. Lingdingyang Bridge is a 580m+1666m+580m three-span full floating suspension bridge, which is one of the largest suspension bridges in the world. Because the bridge deck is up to 90 meters high, when completed, it will be the world's highest bridge in the sea, and its comprehensive scale and difficulty are greater than those of similar projects in the world [2]. The cable tower adopts a portal shape, and its main components include lower tower columns, middle tower columns, upper tower columns, upper beams, middle beams, lower beams, tower crowns, etc. The main structure of the east cable tower is shown in Figure 1. The bottom elevation of the east main tower columns of Lingdingyang Bridge is+0m with a total height of 270m. The design elevation of the upper beams is from +251.826m to +262.500m. See Figure 2 for the structure diagram of the upper beam. The upper beam requires 2248.8m³ C55 marine concrete and 521.1t HRB400 ordinary reinforcement. The design for this height, span and weight of the uppercross beams of the main tower is rare in domestic project, so the process of the design and construction is complicated and difficult [3].

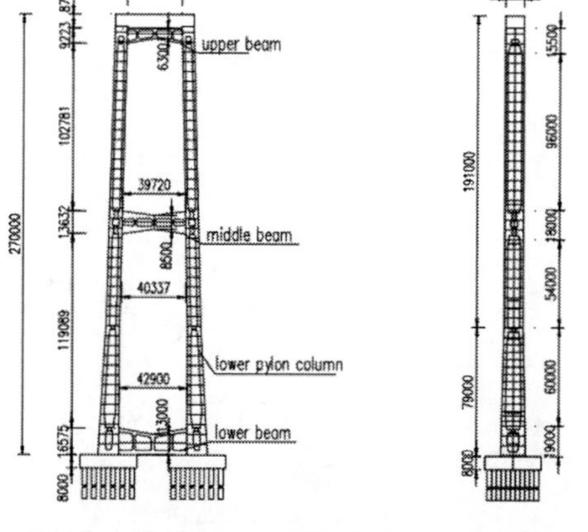

Figure 1. The structure of the east cable tower on Lingdingyang Bridge

* Corresponding author: 307511590@qq.com

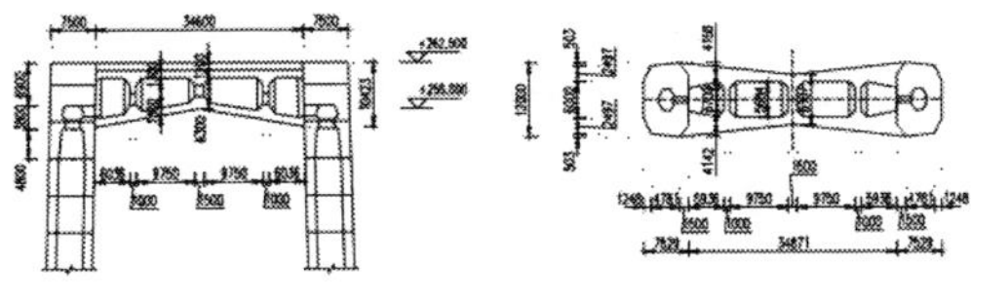

Figure 2. The structure of the upper beam at the east cable tower on Lingdingyang Bridge

2. Design Schemes

The upper beam support of the cable tower is one of the most important temporary structures in the construction. It must have sufficient strength and stiffness, and attention should be paid to the detailed design of the structure. At the same time, the construction method of the support and the safety in the construction process are also very important [4]. Therefore, it is particularly significant to select the design scheme of the upper beam support of the tower and obtain the optimal scheme which is safe, economical and reasonable and convenient for construction.

For the sake of preventing the concrete from producing stable cracks, the upper crossbeam is cast in situ, and it is considered to be cast in two times. In the first layer, the bottom plates, some webs and diaphragms of the upper beam are poured with the pouring height of about 3.30m~6.22m. For the second layer, only the remaining 3m height and 1.2m grooves on both sides are poured. The support only bears the first layer of concrete load with a pouring volume of about 1289m3 and a total weight of about 3351t. The east main tower body is provided with four horizontal braces. The position and structure of the horizontal braces and tower crane buttresses must be considered in combination with the structure type of the upper beam support to avoid layout conflicts. According to the experiences in the similar projects [5-7] and the characteristics of this project, three design schemes are determined.

2.1 Scheme Introduction

The upper beam support structure of Scheme I is mainly composed of support frames, corbels, main beams, trusses, coupling beams, unloading blocks, distribution beams, formworks, etc. The overall structure shows in Figure 3a. Three triangular brackets are symmetrically arranged along both sides of the bridge centerline to form the support platform with a spacing of 3.0m along the bridge. 2HN900× 300 steel is chose for support frames, corbels and coupling beams, the main beams are made of 2HN800 × 300 steel, trusses use HM588 × 300 steel, and the upper distribution beams are laid 20a I-beam according to the spacing of 0.4m. The triangular brackets are connected and fixed with the embedded anchor bars of the tower columns through the bolts at the corbels. The support is about 12m high, 12 embedded parts are set inside the tower bodies, and the total weight of the support is about 600t.

As shown in Figure 3b, the structure of Scheme II consists of steel columns, main beams, trusses, parallel connections, diagonal braces, unloading blocks, distribution beams, formworks, etc. About 20Φ800x10 columns are symmetrically organized along the center line of the bridge in 3 or 4 rows and 6 columns with a height of around 103m, 20 embedded parts of the buttresses are set inside the tower bodies, and the materials of other sections are the same as Scheme I. The total weight is nearly 950t.

The structure in Scheme III is made of steel columns, support frames, corbels, main beams, trusses, parallel connections, diagonal braces, unloading blocks, distribution beams, formworks, etc in Figure 3c. 12Φ800x10 steel columns about 103m high are symmetrically arrayed along the centerline in 3 rows and 4 columns with the same materials of other parts as Scheme I. The structure weighs approximately 950t.

The finite element software is used to model and analyze the three schemes, and the main calculations are shown in Table 1.

Table 1. The main results of design schemes for the upper beam

Design scheme	Vertical displacement (mm)	Shear forces at embedded parts (t)	Horizontal forces at embedded parts (t)
SchemeI	22	embedded parts above: 340 embedded parts below: 440	embedded parts above: 33(tension) embedded parts below: 370 (pressure)
SchemeII	40	Pressures from the steel columns: 250t	
Scheme III	33	embedded parts above: 110 embedded parts below: 265	embedded parts above: 150 (tension) embedded parts below: 270 (pressure)

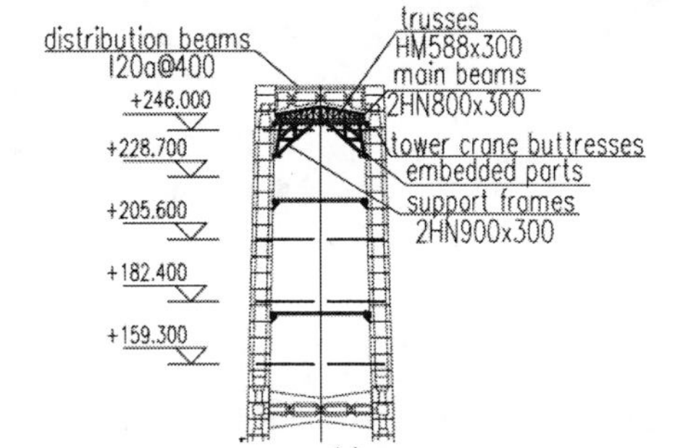

(a)

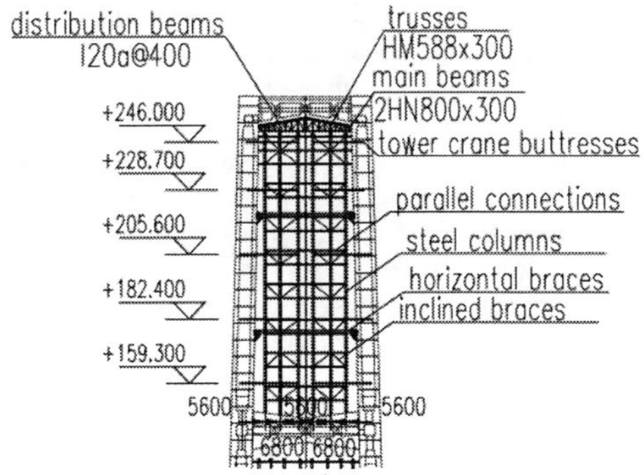

(b)

https://doi.org/10.1051/matecconf/202338001027

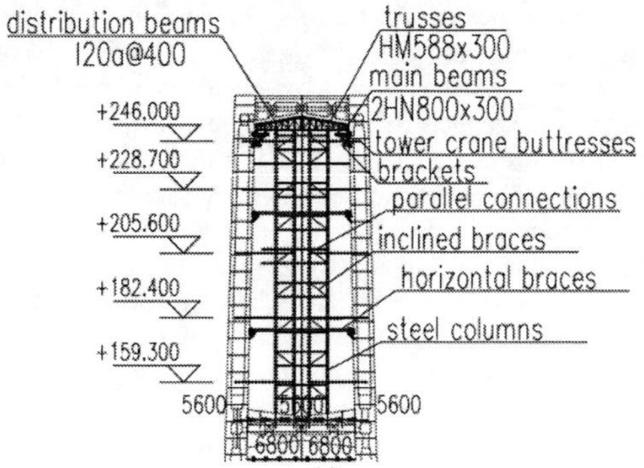

(c)

Figure 3. (a) The elevation of Scheme I (unit: mm), (b) The elevation of Scheme Ⅱ (unit: mm), (c) The elevation of Scheme Ⅲ (unit: mm)

2.2 Scheme Comparisons

The advantages for Scheme I are strong lateral wind resistance, small vertical deformation and low construction cost, while the disadvantages are that the embedded parts of the corbels under the support frame bear too much loads and the weight shared by the support frames on both sides is too large. In addition, it makes the construction difficult and risky, and conflicts with the buttresses of the tower crane.

For the second scheme, the strong point is that it avoids collision with the tower crane buttresses, meanwhile, it can support the horizontal bracing vertically. The weak points are that the column buttresses bear large stress, maximum vertical deformation, superelevation of support, poor lateral wind resistant ability, and high risk of instability under the influence of wind load.

In Scheme III, the good aspects are that the construction and installation are slightly less difficult, at the same time, the bracket can be reused. However, the vertical stiffness between the brackets and the steel columns is greatly different, leading to the uncoordinated vertical deformation. Apart from that, the loads borne by the brackets and the embedded parts at the lower corbel are large, while the bending stress of the truss' lower chord and the vertical bar at the bracket end exceeds the limit stress. Also like Scheme I, the small brackets of the upper beam support collide with the tower crane buttresses.

It can be seen that for the floor-type steel tube brackets such as Scheme II and Scheme III, the height of the steel column is over 100m, and the support is greatly deformed due to the wind load, temperature change and other factors, which will affect the quality of concrete pouring. Meantime the support needs to be connected with the tower columns by setting up buttresses from bottom to top. The space occupied by the support is large, which will affect the layout of the tower crane, tower crane buttresses and other structures. The support weighs heavily and costs much. For Scheme I, it can ensure the quality of concrete pouring for the upper crossbeam, and the area required for the support is also small. The total weight of the support is smaller than that of the steel tube structure, which reduces the steel cost. For the problems existing in the structure, the support can be planned as a parallel arch bracket, besides, brackets and embedded parts can be added to improve resistance at the corbels' embedded parts. Apart from that, the structure could be optimized by strengthening the local position of the truss vertical bar and adjusting the position of the tower crane buttresses to avoid conflicts with the layout. Considering the difficulty of construction, the bracket itself and the bracket embedded parts shall be designed separately, and their angle and position should be arranged during construction before welding to provide the convenience for construction. Scheme I was finally applied to site construction as shown in the figure below.

Figure 4. The construction situation on the spot

3. Conclusions and Suggestions

By comparing the design schemes of three conventional upper beams, the bracket method in the cast-in-place construction is finally determined for the upper beams at the east cable tower of Lingdingyang Bridge on Shenzhen-Zhongyang Link. The design scheme of cast-in-place beam with bracket method provides a reference for the design of the super-height, large-weight and long-span crossbeams at cable tower in complex marine environment, and brings some experience for the construction of cross sea bridges in domestic projects.

(1) The upper arch support can avoid the problem that the structural deformation of the floor-type steel tube bracket is large due to the influence of factors such as wind load and temperature change, which impacts on the concrete pouring quality of the upper beams. It can also optimize the structure, solving the problem of excessive stress on the embedded parts at the corbels and great construction difficulties. It has strong operability and saves costs effectively.

(2) During the construction process, the position and elevation of the buried parts of the brackets' embedded parts should be set accurately to meet the requirements of installation accuracy. All components shall be firmly connected by welding. At the same time, strengthening the on-site inspection and reinforcement work to ensure the safety of the upper beams during the whole construction process.

References

1. Song S Y, GuoJ, Su Q K, etal. Technical Challenges in the Construction of Bridge-Tunnel Sea-Crossing Projects in China[J]. Journal of Zhejiang University Science A: Applied Physics & Engineering , 2020, 21(7): 509-513.
2. XU Guoping, HUANG Qingfei. General Design of Shenzhen-Zhongshan River-crossing Link Project[J]. Tunnel Construction, 2018, 38(04): 627-637.
3. SONG Shenyou, CHEN Weile, JIN Wenliang, et al. Key technologies and challenges of Shenzhong Link[J]. Tunnel Construction, 2020, 40(1):143.
4. Yan Liyuan. Comparison and Choice of the Design Schemes for Brackets at the Top of the Cable Towers of the Self-Anchored Suspension Bridge. Traffic Engineering and Technology for National Defence, 2012, 10(03): 18-20+10.
5. LI Lu-ping. Design and Construction Technology of the Bracket Support of Entablature of the Tower. Value Engineering 2016, 35(16): 98-101.
6. ZHU Peng, ZHOU Xudong. Comparion and Analysis on Falsework Scheme for Upper Cross Beam of Main Tower of Dasha Channel Bridge of the Second Humen Bridge. Guangdong Highway Communications. 2018,44(04): 83-86.
7. Fu Wenxuan, YuanCan, ZhangKe, LiuGe, Huang Xiuping. Key Technology of Design and Construction of Cross-beam Supports for Main Tower of Wufengshan Yangtze River Bridge. Construction Techonology, 2022, 51(06): 19-23+35.

Research on Modal Test for a Double-layered Stiffened Plate in Water

Peng Wang [1], Zhao Li [1], Jiandong Chen [2], Wei Li [3], Hao Zhang [1,*]

[1] Wuhan Second Ship Design and Research Institute, Wuhan 430205, China
[2] Ship Development and Design Center, Wuhan, 430064, China
[3] School of Naval architecture and Ocean Engineering, Huazhong University of Science and Technology, Wuhan 430074, China

Abstract. Madal performance is the basic characteristics of structures, which will help to further understand the vibration of structures. Besides, fluid (e.g. water) coupled with structure will strongly affact the modal performance of structure. To get clear the modal performance, Modal tests are conducted for two-layered plate both in air and in water, and modal frequency and mode shape are given. With comparative analysis of modal frequencies in these two cases, it's shown that influence of fluid on structural vibration is obvious, which will results in significantly reduction of modal frequency, and the maximum reduction is the second mode, and it is more than 42%. But it's also noted that influence of fluid on mode shape is little.

1. Introduction

Plates have wild application in car, ship, aerospace, and so on, it's usually made up of two thin plates and a center core. By filling the empty space in the core with some material (e.g., polyurethane) with good impedance [1], the double-layered plate could get very good property in vibration and noise reduction.

Since 20th century, many researchers have conducted much work on mechanic and vibro-acoustic performance of sandwich structure. Wang et.al.[2] Provided vibration control equation for an orthotropic and infinite plate. Xin et.al.[3] established theoretic mechanic model for finite double-layered plate. Garg et.al.[4] put forward a new high-order theory to analyze the mechanic and free vibration of sandwich beam. Peng et.al.[5] Studied the vibration of sandwich plate with corrugated core using the first shear deformation theory(FSDT). Zhang et.al. [6] study the influence of frame type on dynamic performance of a finite cylindrical shell. It's noted that few work have been done for vibration test for double-layered structures.

In this work, an entity double-layered plate model is constructed, and modal tests in are and water are conducted to get clear the modal performance of double-layered plate.

2. Model test

2.1 Model for test

Experimental model is shown in Fig.1, which is made of upper plate, frame and down plate, the frame is of C section. Due to the need of hanging up of test model, four big steel I shape beam are welded to the four bottom edges of test model.

The up and down plates are of length 690mm and width 690mm, and 3mm thickness, For the frame, the thickness is 2mm, the height is 18.8mm, and the width is 5mm.

Fig. 1 entity model for double-layered plate

2.2 Material parameters

Materials for test model are listed in Tab.1.

Tab. 1 Material parameters

Material	Elastic module (Pa)	Density (g/cm^3)	Poisson's Ratio
Test model	1.93×10^{11}	7.90	0.247
I-shape beam	2.10×10^{11}	7.85	0.300

2.3 Layout of tap Point

Measuring method used in the work is multi-tap points and four fixed measuring points. Tap points are evenly distributed on the face of up plate, and 13 points are planned for both length and width direction of the plate, so the number of tap points is 169. Four measuring points are shown in Fig.2.

* Corresponding author: zhyf1408@126.com

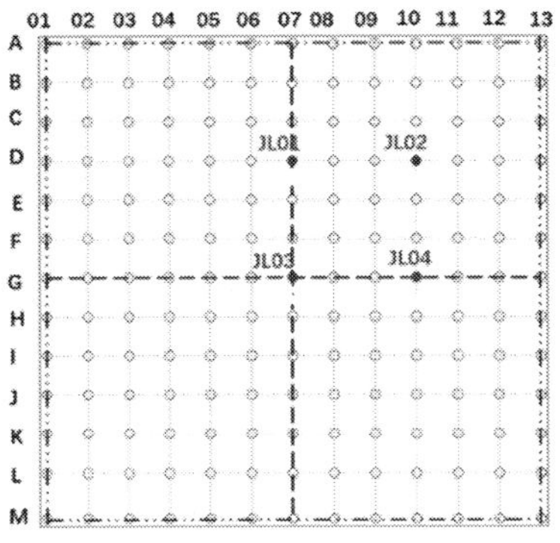

Fig. 2 Location for tap points in modal test

3. Results analysis

3.1 Modal test in air

Modal test for the model in air is firstly carried out. After modal test, the first four modal frequencies and mode shapes are given, as is shown if Fig.3.

In air, the first modal frequency of test model is 306.25Hz, and the mode shape is shown with deformation mainly in the center of plate; The second modal frequency is 434.38Hz, and the mode shape is of 2 half wave in width direction; The third modal frequency is 565.63, and the mode shape is of 3 half waves in width direction.

(a) f=306.25Hz; (b) f=434.38Hz; (c) f=565.63Hz

Fig. 3 Modal test results for double-layered plate in air

3.2 Modal test in water

Then modal test in water is conducted, and the model is hanged up with four spring and cords tied to four corner of big I-shape beams. And the modal frequencies and mode shape are given in Fig.4.

In water, the first modal frequency for test model is 182.81Hz, and the mode shape is of deformation mainly at the center of plate; The second modal frequency is 250Hz, and the mode shape is of 2-half waves in width direction; The third modal frequency is 395.31Hz, and the mode shape is of 3-half waves in width direction.

(a) f=182.81Hz; (b) f=250.00Hz; (c) f=395.31Hz;

Fig. 4 Modal test results for double-layered plate in water

3.3 Comparative analysis

Modal parameters for test model in water are compared with those in the case of "in air", and results are shown in Tab.1

It can be concluded from Tab.1: compared with results in case of "in air", modal frequencies of test model in the case of "In water" are obviously reduced, and the reduction amplitudes of different mode order is different, which means that the influence of fluid-structure interaction on different structural modal frequencies is different, reduction of modal frequency for the first mode is 40.31%, reduction for third mode is 30%, and the second mode is of modal frequency reduction of 42.45%, which is the maximum in first 3 mode order. This phenomenon could be explained as follows, In case of "in water", structure is coupled with fluid, which will generate the additional water mass, so the total mass of fluid-structure model will increase, which will result in the reduction of modal frequencies.

Besides, comparative analysis is also conducted for mode shape in these two cases, and it can be concluded that the change of fluid mterical will make no difference on structural mode shape, despite of some deviations of local points on plates due to measure error.

Tab. 2 Comparative analysis of modal frequency in water and in air (unit: Hz)

Mode order	In air	In water	reduction
1	306.25	182.81	40.31%
2	434.38	250.00	42.45%
3	565.63	395.31	30.11%

4. Conclusions

Modal tests are conducted for two-layered plate both in air and in water, and modal frequency and mode shape are given. For mode shape in two cases, the first mode is of deformation mainly at the center of test plate, and deformation for the second mode is of 2-half waves in width direction, and 3-half waves for the third mode.

With comparative analysis of modal frequencies in these two cases, it's shown that influence of fluid on structural vibration is obvious, which will results in significantly reduction of modal frequency, and the maximum reduction is the second mode, and it is more than 42%. But it's also noted that influence of fluid on mode shape is little.

Conclusions in this work could lay foundation for forced vibration analysis and vibration reduction with core martial in the future.

Acknowledgements

We sincerely thank the financial support from the Hubei Provincial Natural Science Foundation (No. ZRMS2020002231) and the Young Top-notch Talent Cultivation Program of Hubei Province.

References

1. Brooking M A. The performance, safety and production benefits of SPS structures for double hull tankers[C]//The RINA Conference on double Hull Tankers in London, 25th Feb, 2004, 2004.
2. Wang S C, Deng Z X, Shen W D. Sound transmission loss characteristics of unbounded orthotropic sandwich panels in bending vibration considering transverse shear deformation [J]. Composite Structures, 2010, 92: 2885-2889.
3. Xin F X, Lu T J, Chen C Q. Sound transmission through simply supported finite double-panel partitions with enclosed air cavity[J]. Journal of Vibration and Acoustics, 2010, 132: 011008-1-11.
4. Garg A, Chalak H D, Sayyad A S, et al. Novel higher-order zigzag theory for analysis of laminated sandwich beams[J]. Proceedings of the Institution of Mechanical Engineers, Part L: Journal of Materials: Design and Applications, 2021, 235(1): 176-194.
5. Peng L X, Yan S, Mo G, et al. Free vibration analysis of corrugated-core sandwich panel using a meshfree galerkin method based on the first-order shear deformation theory[J]. International Journal of Mechanical Sciences, 2014, 78: 8-18.
6. ZHANG Chao, CHEN Xiao-ning, XU Zhi-ping, Influence of the Frame Shape on the Collision Behavior of the stiffened-Cylindrical Shell. Ship engineering, 2010, 39(1): 23-26 (In Chinese)

Feasibility study on construction of RCC gravity dam under special climatic conditions in Tibet

Ning Pu*, Jingyu Wang, Weibin Zhong, Nan Zheng, Jinming Ren

Powerchina Huadong Engineering Corporation Limited, Hangzhou, China

Abstract. According to the experience in the construction of water conservancy and hydropower projects in the Mainland, RCC dam construction has two main advantages: the rapid construction speed of roller compacted concrete can realize early power generation, and the price of fly ash is much lower than that of cement, which can reduce the project cost. Tibet has special geographical environment and climatic conditions, and generally has the characteristics of "high altitude, low pressure, low temperature, large temperature difference between day and night, and dry climate". Taking the dam-building environment in central Tibet as an example, through investigation and research and analogy of similar projects, this paper analyzes the adaptability of construction of RCC gravity dams in Tibet from the aspects of geographical environment, climatic conditions, material properties, construction progress, and project cost. Adaptability to high altitudes. It provides a reference for choosing a safe, reliable, economical and reasonable dam type in the construction of water conservancy and hydropower projects. It provides reference for selecting safe, reliable, economical and reasonable dam types in water conservancy and hydropower engineering construction.

Key words: High altitude; RCC gravity dam; Alpine climate; Material properties; Adaptability.

1. Introduction

Gravity dams are widely used because of their simple shape, easy flood discharge and adaptability to various foundation conditions. However, concrete dams rely on themselves to maintain stability, so the cross-section is inevitably too large, and the material strength is not fully utilized. Conventional construction methods result in long construction periods and high cost. As a type of concrete gravity dam, roller compacted concrete gravity dam has obvious advantages over normal concrete gravity dam in construction speed and project cost. Since the 1980s, it has developed rapidly. At present, China has become the country with the most RCC gravity dams, and the height of dam construction and the development of dam construction technology are in the forefront of the world [1-2].

Tibet is a rich mine of hydropower in China. Its theoretical reserves of hydropower resources are 201 million kW, accounting for 29% of the total and ranking first in China. The technology exploitable installed capacity is 143 million kW, accounting for 24.9% of the national total, ranking second only to Sichuan Province. Among them, the installed capacity of large-scale hydropower stations in Tibet is 130 million kW, with an annual power generation of about 674 billion kW·h; the installed capacity of medium-sized hydropower stations is 11.72 million kW, with an annual power generation capacity of 62 billion kW·h; the installed capacity of small hydropower stations is 1.18 million kW, with an annual power generation of 6 billion kW·h. According to the development sequence of hydropower resources in China, the future hydropower construction will gradually be transferred to Tibet [3].

According to the experience of china's water conservancy and hydropower engineering construction, RCC dam construction has two main advantages: the rapid construction speed of roller compacted concrete can realize early power generation, and the price of fly ash is much lower than that of cement, which can reduce the project cost. In view of the special geographical environment and climatic conditions, Tibet is generally characterized by "high altitude, low atmospheric pressure, low temperature, large temperature difference between day and night, and dry climate". Whether the long-term experience accumulated in inland water conservancy and hydropower project construction is applicable to Tibet deserves further study and discussion.

Taking the dam-building environment in central Tibet as an example, the author analyzes the adaptability of RCC gravity dam in high altitude area of Tibet from geographical environment, climatic conditions, material properties, construction schedule, cost and other aspects.

* Corresponding author: pu_n@hdec.com

2. Engineering conditions

2.1 Traffic Conditions

(1) Highway. The existing highway network in central Tibet mainly includes G318 National Road, G109 National Road, G312 National Road, S306 Provincial Road, S101 Provincial Road, etc. The road transportation network is relatively developed. From Lhasa City, you can reach Zedang Town via G318 National Highway and S101 Provincial Highway, and then you can reach Bayi Town via S306 Provincial Highway, which is connected to the G318 National Road, and continues eastward to Chengdu (G318 National Road, namely Sichuan-Tibet Line) and Kunming (G318 National Road turns to G214 National Road and G320 National Road, namely Yunnan-Tibet Line).In addition, Xining, Lanzhou and other places can be reached from Lhasa via G109 and G312 National Highways (Qinghai-Tibet Line).
(2) Railway. At present, only the Qinghai-Tibet Railway has been opened in Tibet to the outside world. It is a single-line Class I electrified railway from Xining to Lhasa. The Lari Railway has been opened in Tibet (from Lhasa to Shigatse), and the Lalin Railway (from Lhasa to Linzhi) has also started construction. The materials and equipment can be transported by rail to road to the project site.
(3) Aviation. There is Kongga international airport on the west side of kongga County, Lhasa City, which has opened routes to many major cities in China, such as Beijing, Shanghai, Chengdu and other places.

2.2 Hydrometeorological conditions

The floods of rivers in central Tibet are mainly formed by torrential rain, and the flood and dry flows are very different. The flood season is from June to October, and the dry season is from November to May of the following year.

Central Tibet is mainly located in the climate zone of the Qinghai-Tibet Plateau. Its basic characteristics are low temperature, thin air, dry atmosphere, abnormally strong solar radiation, and complex and diverse climatic characteristics. Taking the measured data of a weather station in central Tibet from 1978 to 2009 as an example, the annual average temperature is 9.3 °C, the average temperature of the coldest month is 0.3 °C (January), and the extreme maximum and minimum temperatures are 32.5 °C and -16.6 °C, respectively.

2.3 Concrete raw material supply status

Concrete raw materials mainly include aggregates, cement and admixtures. Aggregate can be processed with excavated stone or natural sand and gravel near the project area; cement can be supplied from Gaozheng Cement Plant near Lhasa City and Huaxin Cement Plant near Shannan City. The admixture is mainly fly ash, which is not available in Tibet. Qinghai Province, which is closer to Tibet, also has less fly ash resources, while Gansu Province and Ningxia Autonomous Region have a large number of thermal power plants, such as Lanzhou Datang Xigu Power Plant, Yongdeng Liancheng Power Plant, Ningxia Ningwu Power Plant, etc., fly ash can be purchased from the above two places.

3. Adaptability to climatic conditions

The adaptability analysis to climatic conditions is firstly to collect the data of the existing and under-construction projects under similar climatic conditions, summarizes the existing engineering experience, and then analyzes the adaptability of the RCC dam construction in this area through the engineering analogy method. The climatic characteristics of similar projects are shown in Table 1.

Table 1. Similar engineering climatic conditions

project name	Average temperature over the years /°C	Average temperature of the coldest month /°C	climate characteristics	Area	Extreme temperatures /°C		maximum Dam height /m	Completion time
					max	min		
Chonghuer	2.6	-17.3	Cold	Xinjiang	36.6	-45	71	1995
Karasok [4]	2.8	-20.8	Cold	Xinjiang	40.1	-49.8	121.5	1998
Guanyinge	6.2	-14.3	Cold	Liaoning		-32.9	82	2001
Baishi	7.8	-11	Cold	Liaoning		-37	50.3	2008
Longshou	8.5	-15	Cold	Gansu	37.2	-33	80	2009
Luogu	10.9	1.5	Mild	Sichuan	33.1	-20.6	82	2009
Longkaikou	19.8	14.1	Mild	Yunnan	41.8	-2.1	119	2014
Central Tibet	9.3	0.3	Mild	Tibet	32.5	-16.6		

The average temperature of the coldest month measured by a meteorological station in central Tibet is 0.3 °C, which is at the upper level of the projects listed in Table 1. According to the relevant provisions of China's "Code for Design Specification Hydraulic Concrete Structures" and "Design Code for Hydraulic Structures Against Ice and Freezing Action", this area belongs to a mild area[5-6]. In addition, the annual evaporation (2 084.1 mm) and relative humidity (51%) in this area are also comparable to those of similar projects (Luogu: 1805.2 mm, 70%; Longkaikou: 2778.7 mm, 51%). In summary, the RCC gravity dam can adapt to the climatic conditions of the central Tibet.

4. Adaptability of material properties

According to the relevant provisions of the "Technical Specification for Durability on Concrete of Hydraulic Engineering", the frost resistance grade of a RCC gravity dam currently in the design stage in the central Tibet area is C9020F200, and the frost resistance grade of the dam body is C9015F100 and C9020 F100[7]. The results of the RCC mix ratio test are shown in Table 2.

Table 2. Concrete mix ratio of a project in central Tibet for the RCC gravity dam

Design specifications	Sand rate/%	W/B	Water/ (kg•m^{-3})	FA/%	water-reducing admixture/%	air entraining agent /%	Volumetric weight /(kg•m^{-3})	Cement /(kg•m^{-3})	Fly ash /(kg•m^{-3})
Two-graded concrete C_{90}20W8F200	35	0.45	94	50	0.8	10	2370	104	104
Three-graded concrete C_{90}15W6F100,C_{90}20W6F100	30	0.5	88	60	0.8	12	2400	70	106

For RCC dams in severe cold areas in China, such as Longshou in Gansu and Kalasuke in Xinjiang, the minimum temperature is lower, and the required frost resistance grade of external RCC is F300, which is higher than the frost resistance grade of external RCC for projects in central Tibet. The frost resistance grade of the internal concrete is F50~F100, which is similar to the project in central Tibet. The mix ratio of Gansu Longshou and Xinjiang Kalasuke RCC[8] are shown in Table 3.

Table 3. RCC construction mix ratio of Longshou and Kalasuke dam

Design specifications	Sand rate/%	W/ B	Water /(kg•m^{-3})	FA/%	water-reducing admixture/%	air entraining agent /%	Volumetric weight /(kg•m^{-3})	Cement /(kg•m^{-3})	Fly ash /(kg•m^{-3})
Gansu,Longshou	Two-graded concrete C_{90}20W8F300	32	0.43	88	0.9	45	2 400	96	109
	Three-graded concrete C_{90}20W6F100	30	0.48	82	0.9	7	2 400	58	113
Xinjiang,Kelasuke	Two-graded concrete C_{180}20W10F300	34	0.45	98	0.85	12	2 370	131	116
	Three-graded concrete C_{90}20W8F100	27	0.45	90	0.85	8	2 400	140	109
	Three-graded concrete C_{180}15W4F50	28	0.56	90	0.85	6	2 400	61	171

After analysis and comparison, the multi-year average temperature and extreme minimum temperature in central Tibet are higher than those of Longshou and Kalasuk, while the RCC dams of Gansu Longshou and Xinjiang Kalasuk have been in normal operation for many years, indicating that RCC dam construction technology is feasible in cold regions. Combined with the RCC test results of a hydropower station project in the design stage in this area, it shows that the use of RCC dam construction technology in this area is feasible, and there are no restrictive factors in material properties.

5. Adaptability of construction schedule

Central Tibet is located on the Qinghai-Tibet Plateau, with high altitude, large temperature difference between day and night, and low temperature in winter. The project construction needs to overcome unfavorable factors such as plateau hypoxia and cold winter.

RCC dams have the advantages of fast construction speed and flexible construction schedule. According to similar engineering experience, when the monthly average temperature in winter is lower than 0 °C, the construction will be stopped, and polystyrene boards, thermal insulation quilts, etc. are used for winter insulation to ensure the construction quality. Compared with normal concrete dams, under the same construction period, RCC dam can improve construction conditions, reduce the difficulty and cost of temperature control in winter construction, and ensure the construction quality. Therefore, the RCC dam in this area has good adaptability in terms of construction progress.

6. Adaptability of project cost

Based on similar engineering experience, the amount of fly ash in RCC is increased by about 60% on the basis of normal concrete. Since there is no fly ash resource in Tibet, fly ash needs to be purchased from Gansu Province or Ningxia Autonomous Region, and the transportation price is high.

According to the investigation, the ex-factory price of cement in Tibet is about 750 yuan/t, and the freight is about 1.5 yuan/t/km. The ex-factory price of fly ash in

Ningxia Autonomous Region is about 120 yuan /t, and the freight for transporting it by rail to central Tibet is about 750 yuan/t. Preliminary estimates based on the ex-factory price plus freight show that the prices of cement and fly ash arriving at the project are equivalent. The advantage of RCC dam in reducing the project cost through raw materials is not obvious, but there are no constraints on the supply of materials.

7. Conclusions

(1) It is technically feasible to use roller compacted concrete to build dams in central Tibet. The climatic conditions are better than those in Northeast China and Xinjiang. However, it still has the characteristics of high altitude, low pressure, large temperature difference between day and night, and strong sunshine. It is necessary to strengthen research on temperature control and crack prevention measures and interlayer bonding measures for roller compacted concrete, and strictly manage construction quality during the construction process.
(2) Compared with the normal concrete dam, RCC dam has the characteristics of fast construction speed and flexible construction period. It can avoid the influence of cold weather and simplify the temperature control measures in cold season, which is beneficial to ensure the construction quality of the dam body.
(3) Due to the absence of fly ash resources in Tibet, RCC dam has no obvious advantage in reducing project cost through raw materials. However, if admixtures with better mining conditions can be found near the project area to replace fly ash, such as tuff, the advantages of RCC dam in this area will be greatly improved.

References

1. Jiazheng P. Gravity dam design[M]. Beijing: Water Resources and Electric Power Press, 1987: 19.
2. Gongyao S, Sanyi W, Shurong F. High roller compacted concrete gravity dam [M]. Beijing: China Electric Power Press, 2003.
3. Nima DZ, Pian P,Deji ZG. Analysis on the task and existing problems of hydropower resources exploitation in Tibet and their countermeasures [J]. Tibet's Science and Technology, 2012(2): 32.
4. Hui L. Summary of construction technology characteristics and difficulties of Kalasuke Hydraulic Project [J]. Water conservancy and hydropower construction, 2013(6): 17.
5. National Energy Administration. Code for Design Specification Hydraulic Concrete Structures: DL/T5057—2009[S]. Beijing: China Electric Power Press, 2009: 26.
6. National Energy Administration. Design Code for Hydraulic Structures Against Ice and Freezing Action: NB/T35024—2014 [S]. Beijing: China Electric Power Press, 2009: 7.
7. National Energy Administration. Technical Specification for Durability on Concrete of Hydraulic Engineering: DL/T5241—2010[S]. Beijing: China Electric Power Press, 2010: 7.
8. Xiaoli L, Xiaoqi Y, Jinrong H, Zhengbin Z. Research on material properties of roller compacted concrete arch dam of Longshou Hydropower Station[J]. Water Power, 2001(10): 19.

Design and mechanical analysis of shear thickening fluid/polyurethane composite sandwich

Hua Cong, Mingmei Zhao*, Jinqiu Zhang and Yile Liu

Army Armored Forces Academy, Beijing, China

Abstract. In this paper, high density rigid polyurethane foam is used as sandwich skeleton and shear thickening fluid as material core. A shear thickening fluid/polyurethane sandwich structure with light impact resistance was designed and fabricated. High strain rate impact test was carried out. It was found that STF-2/PU reached the peak load of 4978N in 13 ms after receiving 20 J impact energy, and the energy absorption ratio was as high as 43%. The shear thickening fluid/polyurethane honeycomb sandwich foam prepared by secondary foaming process has stable structure and can effectively absorb impact energy to achieve good protection effect.

key word: shear thickening fluid; polyurethane; sandwich structure.

1. Introduction

Shear thickening fluid (STF), as a kind of intelligent material, has a significant increase in apparent viscosity under the condition of applying shear stress or increasing shear rate, which can increase from several times to dozens of times. It is characterized by simple preparation, fast response and reversible repetition [1]. With the deepening of the research, people realize that the shear thickening phenomenon has great application potential in the industrial field. It not only gradually solves various problems caused by shear thickening phenomenon, but also finds that STF has a good application prospect in damping device development [2, 3], vibration reduction, energy absorption and individual protection [4-7] based on its characteristics. Moreover, fruitful achievements have been made in the microscopic mechanism of shear thickening [8-10], rheological properties and their influencing factors [11-14], STF-based development and application [15-17], and other research fields.

Due to the fluidity of shear thickening fluid materials, most of its composite forms are impregnated or coated. In the shear thickening fluid/Kevlar material, the Kevlar fiber is used as the skeleton structure to provide support for the shear thickening fluid. In the face of more complex and violent impact environment, new shear thickening fluid composite forms and skeleton materials need to be proposed. To this, scholars have carried out a lot of research. Albuja et al. [18] filled STF/aramid fabric between two aluminum alloy plates and conducted ballistic firing tests based on NIJ 0108.01 standard. The results show that the target containing STF has better stability after multiple shots. Cwalina et al. [19] replaced the neoprene nylon absorbent layer in the space suit with shear thickening fluid arylon fabric, and conducted impact tests with spherical aluminum projectiles at 4-8 km/s. They found that EVA suits containing STF/fabric provide reasonable micrometeoroid and orbital debris (MMOD) puncture protection while being lighter than standard EVA suits. Pinto et al. [20] filled the shear thickening fluid into the carbon fiber reinforced polymer (CFRP) sandwich plate and found that the energy absorption capacity of STF-CFRP sandwich structure increased with the increase of impact velocity. Jeddi et al. [21] studied the energy absorption characteristics of three-dimensional fabric-core aluminum sandwich plates containing shear thickening fluid (STF) under different loading conditions. The dynamic compression test shows that STF is helpful to improve the impact absorption and compression performance of sandwich plate. Gu et al. [22] filled the shear thickening fluid into the lattice truss sandwich plate (SPLTC-STF) and carried out experiments and numerical simulations at high strain rates. The influence of the shear thickening effect of STF on the dynamic response of SPLTC-STF is predicted, which provides an idea for dynamic energy absorption optimization design of STF-filled sandwich plate under high impact load. The above research shows that the shear thickening fluid can be recombined not only by dipping. When the thickening fluid is shear as the inner core of the sandwich material, the energy absorption capacity of the material can also be improved. On the whole, there are few studies on shear thickening fluid composite sandwich structure at present, most of which are compounded with aluminum plate, steel plate and other materials, and there are some problems in the sealing in the process of composite. On the other hand, the shear thickening fluid composite sandwich structure has a different energy absorption

* Corresponding author: zhaomingmei724@outlook.com

mode than the shear thickening fluid/Kevlar fiber. In the latter, not only the shear thickening behavior can absorb the impact energy, but the use of STF also increases the coupling force and friction between Kevlar yarns, which can also dissipate the impact energy. However, the interaction between non-Newtonian fluid and solid in the shear thickening fluid composite sandwich is mainly the interaction between non-Newtonian fluid and sandwich skeleton, and the interaction between non-Newtonian fluid and sandwich skeleton is worth further study.

Weight, hardness, impact resistance and cost need to be taken into account when choosing the skeleton structure of shear thickening fluid. Polyurethane foams (PU) have significant characteristics such as lightweight, environmental protection, low density, impact resistance, workability and excellent elasticity. Polyurethane foams (PU) are the most widely studied and developed materials in the polymer family [23-25]. By using different polyols and isocyanates as raw materials, polyurethane foams of various physical states such as viscous fluid, low-density soft foam, flexible and semi-flexible foam and hard foam are prepared according to the application requirements of different industries [26]. The density of polyurethane foam can range from 10 kg/m3 to 800 kg/m3 [27]. Its excellent characteristics and low cost make it widely used in automobile protection, furniture manufacturing, sports buffering, transportation, insulation and other fields [28]. Subhash et al. [29] conducted an experimental study on the behavior of polymer PU foam with different porosity under quasi-static and high strain rate loading. Linul [30] et al. studied the mechanical response of rigid polyurethane foams with different densities under impact. In many applications, polyurethane foam is often used as a sandwich structure in aerospace, ships, automobiles, insulated sandwich panels, shipbuilding and construction industries [31]. Liu et al. [32] proposed a composite sandwich structure consisting of fiber-reinforced polymer (FRP), polyurethane foam and ceramic particles, which can effectively protect piers from ship collisions. Chen et al. [33] studied the impact resistance of PU foam sandwich structure under vehicle collision through finite element analysis. Shan [34] proposed a composite protective structure composed of steel plate and polyurethane foam, and found through tests that the modified composite structure had good energy absorption performance. It is not difficult to find that polyurethane foam, a multifunctional material, is suitable for the sandwich skeleton support of shear thickening fluid.

Based on the above discussion, this paper uses high density rigid polyurethane foam as sandwich skeleton and shear thickening fluid as filling material. A shear thickening fluid/polyurethane sandwich structure with light impact resistance was designed and fabricated. The high strain rate impact test was carried out to lay a foundation for the optimization and development of shear thickening fluid sandwich in the future.

2. Shear thickening fluid/polyurethane sandwich structure design

2.1 Raw material

The materials used to prepare STF in this study included nano silica particles (12nm,99.5%, Aladdin Biochemical Technology) and polyethylene glycol (200g/mol, Aladdin Biochemical Technology). Firstly, the mass of each part is calculated according to the concentration of the fluid. Secondly, nano silica particles were dispersed into polyethylene glycol solvent by agitator dispersing instrument (AD500S-H) and stirred for 1 h at 2000 r/min. Since the system is easy to agglomerate, ultrasonic dispersion instrument (GS-010A) was used to vibrate the stirred samples in order to ensure the uniform dispersion of silica particles. Then the samples were placed in a vacuum drying oven for 24 h to remove the bubbles, and finally a stable dispersion system was obtained. The schematic flowchart of preparation is shown in Figure 1. Two shear thickening fluids as shown in Table 1 were prepared by the above process. The rheological characteristics of the two shear thickening fluids are shown in Figure 2.

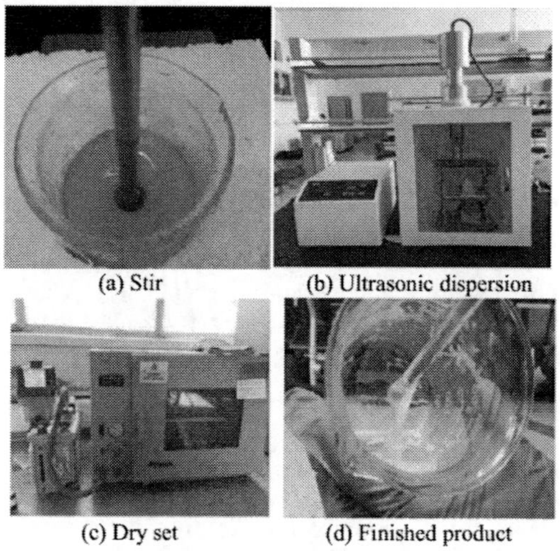

Figure 1. Schematic diagram of preparation of shear thickening fluid

Table 1. Composition of sample of shear thickening fluid

Sample code	Silica particle size (nm)	Silica mass fraction wt.%	Polyethylene glycol Average molecular weight
STF-1(12nm +15wt%SiO_2/PEG200)	12	15	PEG 200
STF-2(12nm +30wt%SiO_2/PEG200)	12	30	PEG 200

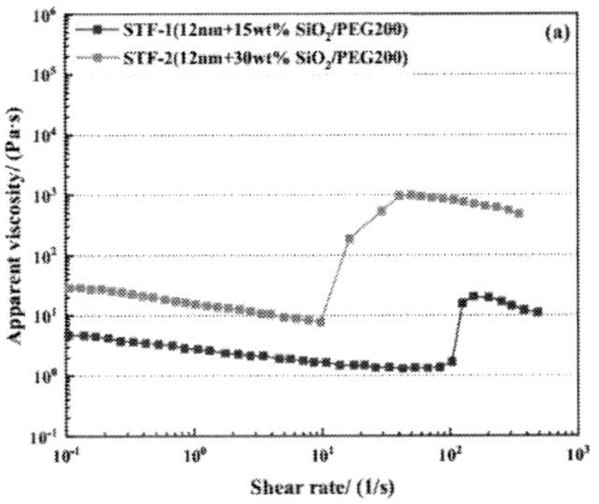

Figure 2. Rheological characteristics of shear thickening fluid

The materials used to prepare polyurethane foam include polyether polyols (Jining Huacai Resin Co., LTD., LDT-02A, density 1.05 g/cm^3) and polyisocyanate (Wanhua Group Co., LTD., 450C, density 1.25 g/cm3). According to the mass ratio of polyether polyol and polyisocyanate at 3:1, the mixture fluid was stirred at 500 r/min by homogenizing stirring dispersion instrument (AD500S-H). The fluid is then mixed and foamed to form polyurethane foam. By controlling the quality of mixed fluid of polyether polyols and polyisocyanate, the density of polyurethane foam is 200 kg/m^3. Scanning electron microscopy was used to observe the prepared polyurethane foam, as shown in Figure 3. As a kind of porous material, polyurethane foam contains closed pores. The compressibility of polyurethane foam was prepared for further testing. The stress and strain of PU material under quasi-static condition was tested by universal testing machine. A number of cubic polyurethane samples (2cm×2cm×2cm) were cut out, and a compression test of 20 mm/min was conducted at room temperature 25℃. The data results of multiple cubic polyurethane samples were averaged, and the results as shown in Figure 4 were obtained.

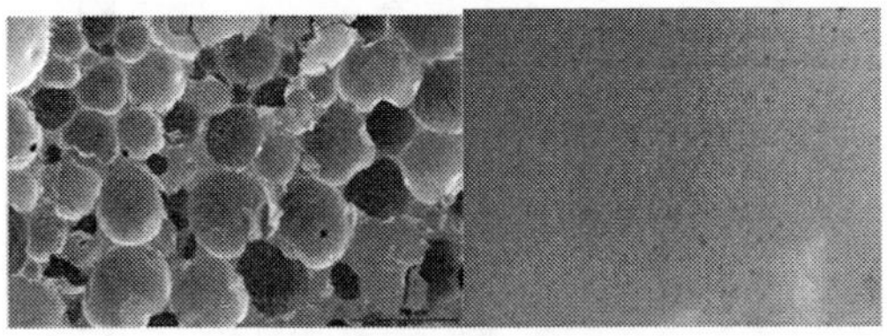

Figure 3. Morphology of polyurethane foam

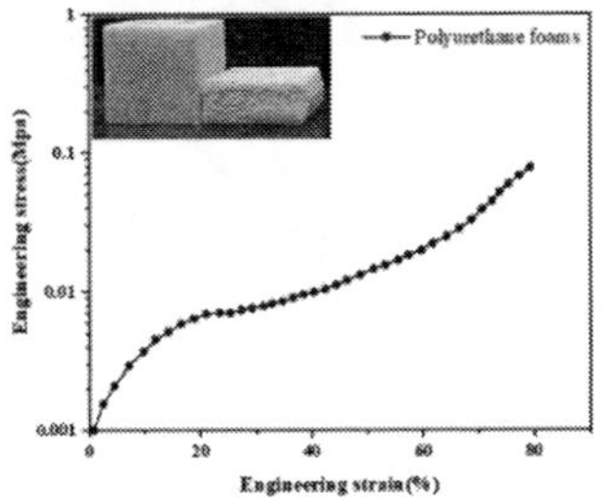

Figure 4. Elastic modulus test of polyurethane foam

2.2 Honeycomb sandwich structure design

The main considerations in the design of sandwich structure are as follows: (1) structural stability. As a fluid, shear thickening fluid needs to be stored stably in a confined space. (2) Light quality. Whether in the field of personal protection, automotive, aviation and industrial manufacturing, it is necessary to reduce the weight of materials as much as possible to achieve the purpose of reducing the load. (3) strong complex integrity. The sandwich structure needs to withstand large deformation in the impact, which requires higher reliability of materials.

Among many sandwich structures, honeycomb sandwich structure is the most widely used and mature [35]. Honeycomb interlayer is produced by imitating the hexagonal honeycomb of bees from the perspective of bionics. This structure has large internal space and strong compressive and bearing capacity [36]. Scholars have carried out a lot of research based on honeycomb sandwich structure. Duan et al. [37]conducted experimental tests and numerical simulation on honeycomb sandwich structure of carbon fiber reinforced polymer, established an oblique impact model, and verified the reliability of the model through the oblique impact test. Rathod et al. [38] studied the mechanical properties and perforation of aramid and aluminum honeycomb sandwich structures under high-speed impact. Through ABAQUS/Explicit modeling, the influence of different geometric parameters on the impact response of sandwich structure was studied. Huang et al. [39] filled honeycomb sandwich with magnetic shear thickening gel, designed and optimized the wave-absorbing honeycomb sandwich structure (MAHSS) by simulating the content of carbonyl iron particles (FCIP) and the properties of magnetic shear thickening gel. Inspired by leaf texture, Lv et al. [40]innovatively added rectangular mesh to honeycomb to form a mesh-honeycomb hybrid sandwich. The low speed impact response of grid-honeycomb composite sandwich structures is studied by means of experiments and numerical simulation.

All the above studies indicate that honeycomb sandwich structure can provide stable structure during dynamic impact process. Therefore, in this study, the polyurethane foam was designed as a honeycomb sandwich structure. Considering the operability of the actual foaming process and preparation process, the honeycomb sandwich structure is designed as shown in Figure 5. The honeycomb structure consists of regular hexagons. The yellow part is polyurethane PU foam, and the hexagonal side length is 2 cm. The blue part is the shear thickening fluid filling part. The side length is 1.5 cm. The overall thickness of the sandwich is 2 cm, and the thickness of the shear thickening fluid is 1 cm.

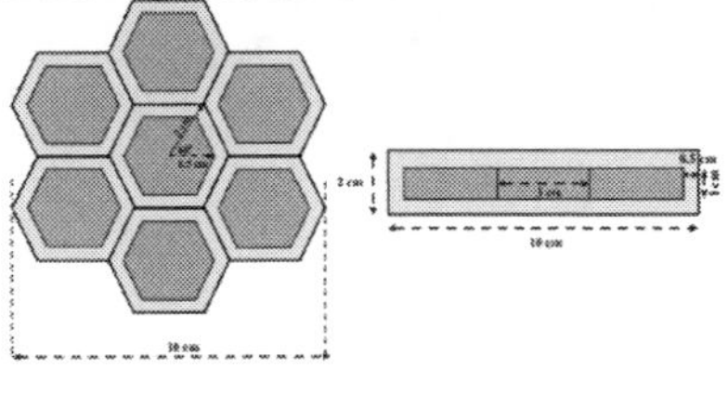

Figure 5. Honeycomb sandwich structure

3. Shear thickening fluid/polyurethane interlayer synthesis

Shear thickening fluid/polyurethane sandwich structure was prepared by secondary foaming process. As shown in Figure 6(a), polyether polyols and polyisocyanate were mixed, quickly stirred, and quickly poured into the honeycomb mold. After the formation of polyurethane foam, foaming was completed in about 25 minutes. According to the thickness setting of the sandwich structure, the thickness of the excess polyurethane is removed to complete the lower half of the STF/PU sandwich, as shown in Figure 6(b). Secondly, pour the prepared shear thickening fluid into the honeycomb space under the same process, as shown in FIG. 6(c) and (d). Then the polyurethane foam loaded with STF fluid was placed into the mold again for secondary foaming and excess thickness was removed, as shown in Figure 6(e). Finally, a cellular polyurethane sandwich foam containing shear thickening fluid is formed, as shown in Figure 6(f). The overall production process and technology is shown in Figure 6.

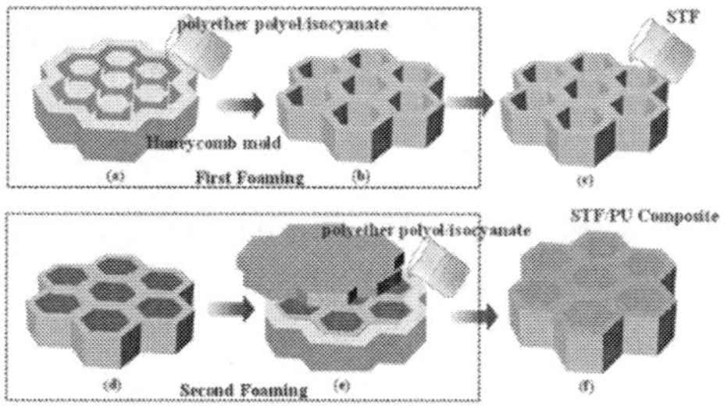

Figure 6. Production process of shear thickening fluid/polyurethane composite sandwich

4. Mechanical characteristics analysis of shear thickening fluid/polyurethane sandwich structure

Considering the small volume and elastic modulus of the shear thickening fluid/polyurethane foam, in order to make the composite sandwich under uniform stress, DYNATUP 8200 drop weight impact testing machine was used in the mechanical test, and a disc-shaped plate with a diameter of 70 mm and a mass of 5 kg was used as the impact object, as shown in Figure 7. By changing the height of the impact plate, the impact kinetic energy of the drop hammer plate is adjusted. The load curves and energy absorption of STF-1/PU and STF-2/PU sandwich structures were tested under the impact of 10J, 15J and 20J respectively (corresponding impact velocities of 2 m/s, 2.45 m/s and 2.83 m/s). In order to reduce sample error, three samples of each type of STF/PU interlayer were tested with an average value.

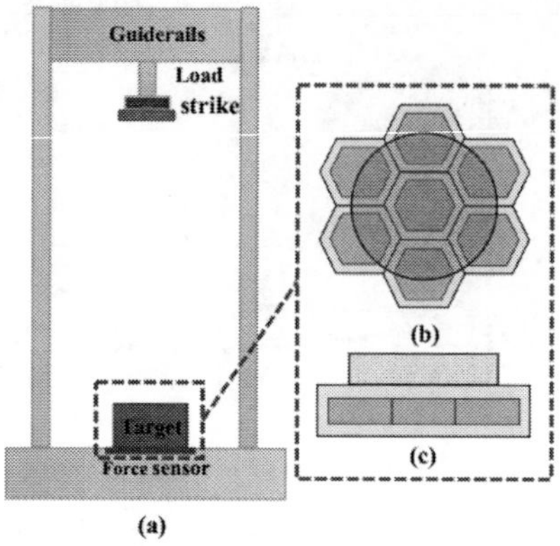

Figure 7. Schematic diagram of shear thickening fluid/polyurethane impact testing equipment

The test results of shear thickening fluid/polyurethane foam under the impact of 10J, 15J and 20J are shown in Figure 8 and Figure 9. Where, (a) is the relationship curve between load force and time detected by the sensor under the STF/PU composite sandwich in the impact process. (b) is the relationship between deformation depth and energy absorption of STF/PU composite sandwich with impact deformation on the sandwich surface after the disc drop hammer contacts the surface.

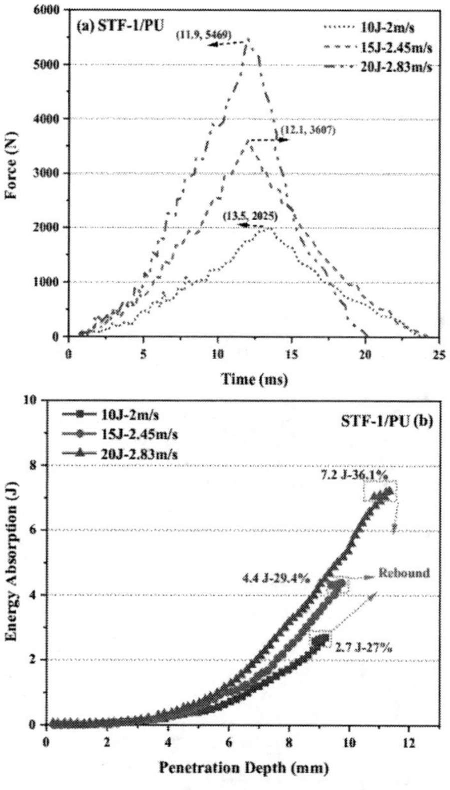

Figure 8. STF-1/PU impact test result
(a) Time-load; (b) Distortion-energy absorption

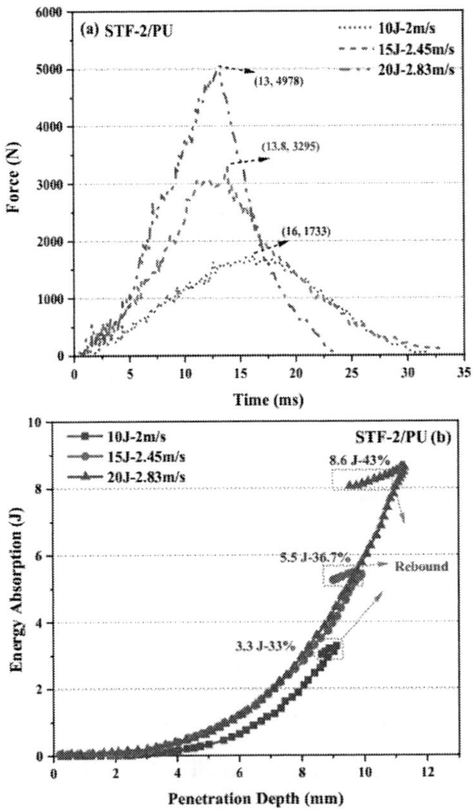

Figure 9. STF-2/PU impact test result
(a) Time-load; (b) Distortion-energy absorption

According to the law of conservation of energy, the expression of energy absorption in the impact process is shown as follows. Where ma is the weight of the drop weight, h is the height of the free fall, F is the load force recorded by the sensor below the sandwich, and z is the penetration depth.

$$E_{Absorption} = m_a gh - \int Fdz \qquad (1)$$

According to Figure 8 and 9, as the impact energy increases, the peak load of the same STF/PU sandwich increases, the reaction time shortens, and the corresponding absorbed energy increases. Among them, STF-1/PU reached the peak load of 5469N in 11.9ms after receiving 20J impact energy, and the energy absorption ratio was as high as 36.1%. After receiving 20J of impact energy, STF-2/PU reaches the peak load of 4978N in 13 ms, and absorbs 8.6J of energy, with energy absorption ratio up to 43%. Meanwhile, it is found in Figure 8(b) and 9(b) that when the absorbed energy reaches the highest value, the penetration depth decreases. This indicates that with the continuous compression of the drop weight, the STF/PU composite sandwich appears springback phenomenon. Consistent PU material was considered in the two tests, indicating that the difference in shear thickening fluid caused the energy absorption and springback of sandwich structures. According to Figure 2, compared with STF-2 fluid, STF-1 with a mass fraction of 15% has weak thickening effect and low maximum viscosity. Therefore, at the same impact velocity, the apparent viscosity of STF-1 in PU interlayer is lower than that of STF-2/PU, and the impact buffering effect is poor.

5. Conclusion

As a typical non - Newtonian fluid, STF has been widely used in the field of shock resistance. But the action mechanism of the composite sandwich structure still needs to be further explored. In this paper, high density rigid polyurethane foam as sandwich skeleton, shear thickening fluid as the material core, design and manufacture a lightweight impact resistant shear thickening fluid/polyurethane sandwich structure. High strain rate impact test was carried out.

It was found that the shear thickening fluid/polyurethane honeycomb sandwich foam prepared by secondary foaming process has stable structure. With the increase of impact energy, the peak load at the bottom of STF/PU honeycomb sandwich increases, the reaction time shortens, and the corresponding absorbed energy increases. Among them, STF-1/PU reached the peak load of 5469N in 11.9ms after receiving 20J impact energy, and the energy absorption ratio was as high as 36.1%. After receiving 20J of impact energy, STF-2/PU reaches the peak load of 4978N in 13 ms, and absorbs 8.6J of energy, with energy absorption ratio up to 43%. It can be seen that the shear thickening fluid/polyurethane sandwich structure can effectively absorb impact energy and achieve good protection effect.

Acknowledgments

This work was funded by the National Natural Science Foundation of China [No. 51605490].

References

1. S. Gürgen, M.C. Kuşhan, W. Li, Shear thickening fluids in protective applications: A review, Progress in Polymer Science, 75 (2017) 48-72.
2. B. Liu, C. Du, H. Deng, Z. Fan, J. Zhang, F. Zeng, Y. Fu, X. Gong, Mechanical properties of magneto-sensitive shear thickening fluid absorber and application potential in a vehicle, Composites Part A: Applied Science and Manufacturing, 154 (2022).
3. Q. Zhao, J. Yuan, H. Jiang, H. Yao, B. Wen, Vibration control of a rotor system by shear thickening fluid dampers, Journal of Sound and Vibration, 494 (2021) 115883.
4. S. Zhang, S. Wang, Y. Wang, X. Fan, L. Ding, S. Xuan, X. Gong, Conductive shear thickening gel/polyurethane sponge: A flexible human motion detection sensor with excellent safeguarding performance, Composites Part A: Applied Science and Manufacturing, 112 (2018) 197-206.
5. S. Sen, N.B. Jamal M, A. Shaw, A. Deb, Numerical investigation of ballistic performance of shear

thickening fluid (STF)-Kevlar composite, International Journal of Mechanical Sciences, 164 (2019).

6. T.F. Santos, C.M.S. Santos, M.S. Aquino, D. Ionesi, J.I. Medeiros, Influence of silane coupling agent on shear thickening fluids (STF) for personal protection, Journal of Materials Research and Technology, 8 (2019) 4032-4039.
7. U. Mawkhlieng, A. Majumdar, Deconstructing the role of shear thickening fluid in enhancing the impact resistance of high-performance fabrics, Composites Part B: Engineering, 175 (2019).
8. S. Cao, Q. He, H. Pang, K. Chen, W. Jiang, X. Gong, Stress relaxation in the transition from shear thinning to shear jamming in shear thickening fluid, Smart Materials and Structures, 27 (2018).
9. E.K. Hadde, J. Chen, Shear and extensional rheological characterization of thickened fluid for dysphagia management, Journal of Food Engineering, 245 (2019) 18-23.
10. M. Liu, W. Jian, S. Wang, S. Xuan, L. Bai, M. Sang, X. Gong, Shear thickening fluid with tunable structural colors, Smart Materials and Structures, 27 (2018).
11. K. Chen, Y. Wang, S. Xuan, X. Gong, A hybrid molecular dynamics study on the non-Newtonian rheological behaviors of shear thickening fluid, J Colloid Interface Sci, 497 (2017) 378-384.
12. J. Ge, Z. Tan, W. Li, H. Zhang, The rheological properties of shear thickening fluid reinforced with SiC nanowires, Results in Physics, 7 (2017) 3369-3372.
13. X. Gong, Q. Chen, M. Liu, S. Cao, S. Xuan, W. Jiang, Squeeze flow behavior of shear thickening fluid under constant volume, Smart Materials and Structures, 26 (2017).
14. S. Gürgen, W. Li, M.C. Kuşhan, The rheology of shear thickening fluids with various ceramic particle additives, Materials & Design, 104 (2016) 312-319.
15. S. Gürgen, M.C. Kuşhan, The ballistic performance of aramid based fabrics impregnated with multi-phase shear thickening fluids, Polymer Testing, 64 (2017) 296-306.
16. S. Gürgen, M.A. Sofuoğlu, Experimental investigation on vibration characteristics of shear thickening fluid filled CFRP tubes, Composite Structures, 226 (2019).
17. A. Khodadadi, G. Liaghat, A. Taherzadeh-Fard, D. Shahgholian-Ghahfarokhi, Impact characteristics of soft composites using shear thickening fluid and natural rubber–A review of current status, Composite Structures, 271 (2021).
18. E.H. Albuja, J.A. Szpunar, A.G. Odeshi, Ballistic impact response of laminated hybrid materials made of 5086-H32 aluminum alloy, epoxy and Kevlar fabrics impregnated with shear thickening fluid, Composites Part A Applied Science and Manufacturing, (2016) 54-65.
19. C.D. Cwalina, R.D. Dombrowski, C.J. Mccutcheon, E.L. Christiansen, N.J. Wagner, MMOD Puncture Resistance of EVA Suits with Shear Thickening Fluid (STF) – Armortm Absorber Layers, Procedia Engineering, 103 (2015) 97-104.
20. F. Pinto, M. Meo, Design and Manufacturing of a Novel Shear Thickening Fluid Composite (STFC) with Enhanced out-of-Plane Properties and Damage Suppression, Applied Composite Materials, (2017).
21. M. Jeddi, M. Yazdani, H. Hasan-nezhad, Energy absorption characteristics of aluminum sandwich panels with Shear Thickening Fluid (STF) filled 3D fabric cores under dynamic loading conditions, Thin-Walled Structures, 168 (2021).
22. Z.P. Gu, X.Q. Wu, Q.M. Li, Q.Y. Yin, C.G. Huang, Dynamic compressive behaviour of sandwich panels with lattice truss core filled by shear thickening fluid, International Journal of Impact Engineering, 143 (2020) 103616.
23. V.A. Lemos, M.S. Santos, E.S. Santos, M. Santos, W. Santos, A.S. Souza, D. Jesus, C. Virgens, M.S. Carvalho, N. Oleszczuk, Application of polyurethane foam as a sorbent for trace metal pre-concentration — A review, Spectrochimica Acta Part B Atomic Spectroscopy, 62 (2007) 4-12.
24. Q. Zhang, X. Yu, F. Scarpa, D. Barton, Y. Zhu, Z.Q. Lang, D. Zhang, A dynamic poroelastic model for auxetic polyurethane foams involving viscoelasticity and pneumatic damping effects in the linear regime, Mechanical Systems & Signal Processing, (2022) 179.
25. Stirna, Beverte, Yakushin, Cabulis, Mechanical properties of rigid polyurethane foams at room and cryogenic temperatures, J CELL PLAST, (2011).
26. A. Hakim, M. Nassar, A. Emam, M. Sultan, Preparation and characterization of rigid polyurethane foam prepared from sugar-cane bagasse polyol, Materials Chemistry & Physics, 129 (2011) 301-307.
27. F. Saint-Michel, L. Chazeau, J.Y. Cavaille, Mechanical properties of high density polyurethane foams: I. Effect of the density, Composites Science & Technology, 66 (2006) 2700-2708.
28. Singh, Harpal, Jain, A., K., Ignition, combustion, toxicity, and fire retardancy of polyurethane foams: A comprehensive review, Journal of Applied Polymer Science, 111 (2008) NA-NA.
29. G. Subhash, Q. Liu, X.L. Gao, Quasistatic and high strain rate uniaxial compressive response of polymeric structural foams.
30. E. Linul, L. Marsavina, D. Apostol, M.C. Dan, T. Sadowski, Effect of Density, Loading Rate, Material Orientation and Temperature on Dynamic Compression Behavior of Rigid PUR Foams, in: 10th International Conference on Foam Materials &Technology, 2012.
31. D. Grzda, G. Wgrzyk, M. Leszczyńska, L. Szczepkowski, M. Gloc, J. Ryszkowska, Viscoelastic

Polyurethane Foams for Use as Auxiliary Materials in Orthopedics, Materials, 15 (2022) 133-.

32. L. Zhu, W. Liu, H. Fang, J. Chen, Y. Zhuang, J. Han, Design and simulation of innovative foam-filled Lattice Composite Bumper System for bridge protection in ship collisions, Composites Part B: Engineering, 157 (2019) 24-35.
33. T. Liu, L. Chen, Numerical Simulation of Vehicle Collision with Reinforced Concrete Piers Protected by FRP-Foam Composites, in: Structures Congress 2019, 2019.
34. C. Shan, Analysis of collision performance of anticollision box made of steel–polyurethane sandwich plates, Journal of Constructional Steel Research, 175 (2020) 106357.
35. Y. Zhang, X. Liu, Y. Zhou, Y. Shi, Uniaxial compressive performance of an aramid and aluminum honeycomb sandwich structure, Ocean Engineering, 270 (2023) 113676.
36. N. Khaire, G. Tiwari, V. Patel, M.A. Iqbal, Assessment of the ballistic response of honeycomb sandwich structures subjected to offset and normal impact, Defence Technology, (2023).
37. Y. Duan, Z. Cui, X. Xie, Y. Tie, T. Zou, T. Wang, Mechanical characteristics of composite honeycomb sandwich structures under oblique impact, Theoretical and Applied Mechanics Letters, 12 (2022) 100379.
38. S. Rathod, N. Khaire, G. Tiwari, A comparative study on the ballistic performance of aramid and aluminum honeycomb sandwich structures, Composite Structures, 299 (2022) 116048.
39. S. Huang, Y. Liu, K. Wen, X. Su, C. Liang, H. Duan, G. Zhao, Optimization design of a novel microwave absorbing honeycomb sandwich structure filled with magnetic shear-stiffening gel, Composites Science and Technology, 232 (2023) 109883.
40. H. Lv, S. Shi, B. Chen, J. Ma, Z. Sun, Low-velocity impact response of composite sandwich structure with grid–honeycomb hybrid core, International Journal of Mechanical Sciences, 246 (2023) 108149.

AUTHOR INDEX

EDP Sciences – Web of Conferences
17, Avenue du Hoggar
Parc d'Activité de Courtabouf
BP 112
F-91944 Les Ulis Cedex A
France

ISBN 978-1-7138-7341-9

CURRAN ASSOCIATES INC.
proceedings.com

International Conference on Renewable Energy and Environment Engineering (REEE 2018)

E3S Web of Conferences Volume 80 (2019)

Paris, France
29 - 31 October 2018

edp sciences